BIOTECHNOLOGICAL INVENTIONS

*For Greta, George, Gerard and Pauline
not forgetting Fionnuala, Mary and Maurice*

Biotechnological Inventions
Moral Restraints and Patent Law
Revised Edition

OLIVER MILLS
National University of Ireland, Galway

ASHGATE

Published by
Ashgate Publishing Limited
Wey Court East
Union Road
Farnham
Surrey, GU9 7PT
England

Ashgate Publishing Company
Suite 420
101 Cherry Street
Burlington
VT 05401-4405
USA

www.ashgate.com

British Library Cataloguing in Publication Data
Mills, Oliver.
    Biotechnological inventions : moral restraints and patent
    law. -- 2nd ed.
    1. Biotechnology--Patents. 2. Patent laws and
    legislation--Europe. 3. Patent laws and legislation--
    United States. 4. Biotechnology industries--Law and
    legislation--Europe. 5. Biotechnology industries--Law and
    legislation--United States.
    I. Title
    346'.0486-dc22

Library of Congress Cataloging-in-Publication Data
Mills, Oliver.
    Biotechnological inventions : moral restraints and patent law / by Oliver Mills.
        p. cm.
    Includes bibliographical references and index.
    ISBN 978-0-7546-7774-1 (hardback) -- ISBN 978-0-7546-9524-0 (ebook)
    1. Biotechnology--Patents. 2. Patent laws and legislation--Europe. 3. Patent
laws and legislation--United States. 4. Biotechnology industries--Law and legislation--
Europe. 5. Biotechnology industries--Law and legislation--United States. I. Title.
    K1519.B54M55 2010
    346.04'86--dc22

2010005660

ISBN 9780754677741 (hbk)
ISBN 9780754695240 (ebk)

Printed and bound in Great Britain by
MPG Books Group, UK

Contents

Table of Cases

Table of Statutes

Table of Conventions

Preface to the Revised Edition

The biotechnology industry is at the cutting edge of scientific research and innovation. The result is that communities, human and non-human, stand on the threshold of an extraordinary revolution that will have profound effects on their relationship with other creatures and the environment. This has resulted in much political debate because countries have fundamentally differing views about biotechnology. In addition, there are divergent views concerning the legal, moral, ethical and social problems connected with the technology. In order to orchestrate an informed debate among the public about the practice and application of biotechnology, scientists and non-scientists alike must work together in an effort to understand what biotechnology can and cannot do. While scientists are capable of identifying the potential advances in the fields of healthcare, agriculture and the food industry that might flow from the commercial application of the technology, only the public can decide their degree of importance and assess the benefits, hazards and impact. Such fundamental decisions can be made only within a properly constituted framework. Law can provide the mechanisms that facilitate the taking of necessary decisions about the nature and direction of both research and policy. In this regard, there are moral objections both to the technology itself and the patenting of biotechnological inventions. The patenting process stands at the confluence of science and technology, on the one hand, and law on the other. In several respects, the intertwining of two disciplines creates tensions not only at the national, but also the international, level. Problems arise because there are differences in matters such as: concepts justifying patent protection; patent law as a matter of economic policy; and patent law as an integral part of the general legal system of the country concerned.

Biotechnology has always been part of our heritage. However, it has been of special concern only since the novel use of organisms in the context of DNA structure was discovered. While developed initially in the area of lower organisms, modern genetic technologies are increasingly applied to more complex biological entities, giving rise to previously unknown, or little-known, concerns of a moral nature.

The advent of biotechnology has posed significant challenges for patent law. As in any legal system there are areas of contention and uncertainty as to the application of legal provisions to particular fact situations. The nature of the subject-matter gives rise to complex conceptual, theoretical and moral questions, particularly in regard to the application of exclusions to patentability of biotechnological inventions.

In its first edition, published in 2005, this book outlined the moral debate surrounding biotechnology and the patenting of biotechnological inventions. The aims of the European Patent Convention were examined and showed that a 'light' moral regime was intended. This compared favourably with the approach to the question of morality in the United States. The book also examined European Patent Office jurisprudence in the light of the policies underlying the exceptions to patentability; in regard to UK Patent Office jurisprudence the cases demonstrated that sufficient difficulties arise when traditional substantive patent law criteria are applied to new technologies without additional assessment of moral concerns. This was compared with protection available to innovators in the biological field in the United States. The Biotechnology Directive 1998 was outlined and assessed in order to determine the extent to which it allows the European Patent Office and national patent-granting authorities to interpret in the same way those provisions of patent law that apply to biotechnological inventions. In addition, the book examined why plant variety rights systems form a part of the critique of the Directive. In European states, developments showed an attempt by the European Union and European Patent Office to confine morality to commercial situations and to bring the European Patent Convention within the confines of the EU.

In such a fast-moving field as biotechnology there has been a plethora of new published material. Much of this emphasizes the higher public profile of debates associated with the technology and includes ethical and privacy issues surrounding the development of bio-banking, including cord blood banking and DNA profiling, and issues surrounding new therapeutic applications in regenerative medicines. Such subjects are undoubtedly interesting but are clearly outside the scope of this work, which is concerned with the legal framework associated with biotechnological inventions.

While there have been no radical changes to the law since publication of the first edition in 2005, there has been a steady accretion of the law. In this regard, the revised edition outlines developments and continues with a focused discussion of specific legal provisions. In particular, the Revised European Patent Convention 2000 (in force 2007) is important. It now incorporates the Biotechnology Directive 1998, and in this regard, Rule 28 EPC 2000 and Rule 29 EPC 2000 are examined to determine the extent to which harmonization of law has been facilitated. Public policy, and changes relating to Article 52(4) EPC 1973, now appearing as the new Article 53(c) EPC 2000, are examined. The revised edition reviews case law developments and examines, in particular, the application of Article 53(a) in relation to stem cells and human genetics. Additionally, new cases have been included to assess whether or not consistency of practice has been achieved by the courts in respect of more traditional patentability criteria. Monitored also through case law developments is how patent law reform in Europe, through the European Patent Office, is expanding; in contrast, patent law reform in the United States, through the Patent and Trademark Office, may now be curtailed. Also discussed are the most recent European Commission communications which highlight the importance of the link between patents and innovation. The European Commission

Proposal for a Regulation is aimed at creating a new unitary industrial property right, the Community patent. This, the Commission believes, is essential for eliminating the distortion of competition resulting from the territorial nature of national protection rights.

Acknowledgements

I wish to thank Ashgate Publishing Ltd for agreeing to publish the revised edition of this book. In particular, Mr John Irwin, Consultant Publisher, and Ms Alison Kirk, the Law Publisher, without whose help this work would not have been completed.

Readers are advised that any shortcomings in the revised edition are solely my responsibility and that no liability will be accepted in respect of any reliance thereon.

List of Abbreviations

AIPPI	International Association for the Protection of Industrial Property
ASSINEL	International Association of Plant Breeders for the Protection of Plant Varieties
BIRPI	Bureaux internationaux réunis pour la protection de la propriété intellectuelle
CAFC	Court of Appeals for the Federal Circuit
CCPA	Court of Customs and Patent Appeals
CPVR	Community Plant Variety Rights 1994
ECHR	European Convention on Human Rights
ECJ	European Court of Justice
EEC	European Economic Community
EFTA	European Free Trade Area
EGE	European Group on Ethics in Science and New Technologies
EPC	European Patent Convention
EPO	European Patent Office
EPOR	European Patent Office Reports
EU	European Union
FSR	Fleet Street Reports
GAEIB	Group of Advisors on Ethical Implications of Biotechnology
OJ	*Official Journal* of the European Communities
PCT	Patent Cooperation Treaty 1970
PIP	1883 Paris Convention for the Protection of Industrial Property
PPA	1930 Plant Patent Act
PTO	US Patent and Trademark Office
PVPA	1970 Plant Variety Protection Act
RPC	Reports on Patent Cases
TRIPS	1994 Agreement on Trade-Related Aspects of Intellectual Property Rights
UPOV 1961	International Union for the Protection of New Varieties of Plants 1961 Act
UPOV 1991	International Union for the Protection of New Varieties of Plants 1991 Act
USC	United States Code
USPQ	United States Patent Quarterly
WIPO	World Intellectual Property Organization

Chapter 1
Biotechnological Inventions: The Moral Debate

Introduction

The biotechnology industry is at the cutting edge of scientific research and innovation. As a scientific art and commercial process it is problematic not merely for scientists and economists but also for lawyers and the general public. Problems for research scientists include how to discover and examine the workings of nature. One of the major challenges for economists lies in learning how to cope with the growth of new industries. In this regard, the economic role of patents is, arguably, more important than ever before, and still, patent systems are only marginally influenced by economists.[1] For lawyers, the fast-moving pace of biotechnology means it is necessary to keep matters under review and to provide an appropriate legislative response. In the interests of the public, the actual and potential impact of the technology, in particular in the domain of health care, agriculture and the environment, needs to be addressed more fully. In this regard, private companies hoping to develop the huge potential for commercial applications must look beyond the technology and into the future if the science is to be brought successfully to market.[2] It is equally imperative to ensure that publicly funded research, where commercialized, results in adequate, and sufficient, public benefit. In terms of regulation, the rate of development of biotechnology means that situations and problems are likely to precede primary legislation and, indeed, case law.[3] Because of this it is necessary to be flexible and to keep under review the development of scientific possibilities, in the context of public understanding, so that as far as possible issues are anticipated, and research and development, and legislation, are conducted with as full as possible an appreciation of the consequences.[4]

In order to orchestrate an informed debate about the practice and applications of biotechnology, the non-scientist public, including lawyers, must understand what,

---

1 Dominique Guellec and Bruno van Pottelsberghe de la Potterie, *The Economics of the European Patent System* (2007) Oxford University Press, UK.

2 See generally Martin Austin, *Business Development for the Biotechnology and Pharmaceutical Industry* (2008) Gower Publishing, UK.

3 See generally R Brownsword, W R Cornish and M Llewelyn (eds), Editors' Introduction: 'Human Genetics and the Law: Regulating a Revolution', 1998 vol. 61 *Modern Law Review* 593–97 at 595.

4 Ibid.

technically speaking, biotechnology can and cannot do.[5] Scientists are capable of identifying the potential advances in the fields of health care, agriculture and the environment that might flow from the commercial applications of biotechnology.[6] However, they are no better equipped than anyone else to identify the social and moral issues that may result from their use. Only society can decide the degree of importance to be attached to these benefits, hazards and impacts. Such fundamental decisions within an area of inherent uncertainty can be made only within a properly constituted framework. Law can provide the mechanisms that facilitate the taking of necessary decisions about the nature and direction of both research and policy. In this sense, besides being instrumental in the policy process, law promotes legitimacy and accountability.

The result of not having a sufficiently constituted framework is that the legal and moral boundaries of biotechnology are unclear to the patent lawyer. Indeed, if the appropriate application and development of legal terms to scientific practices is problematic for those directly involved, how are the concerns of the general public to be addressed?

There are moral objections both to patenting biotechnological inventions and to the technology itself. In this regard, defining morality is one of the key problems. Morality has been described as the system of norms, values and ideals considered universally valid and important to serve as guidelines for action.[7] A moral law can be thought of as a law with no sanction for its transgression other than in conscience. However, in terms of law, where certainty and clarity are foremost (and punishment for breach ensues), such a general definition is too broad to have any significant meaning. In the context of morality the extent to which the patent system (the main features of which are summarized briefly hereafter), in particular in Europe, either initially under the European Patent Convention 1973 (henceforth EPC) and later the Directive on the Legal Protection of Biotechnological Inventions 1998 (henceforth the Directive), and in the United States (henceforth US), is designed, or, indeed, even appropriate to accommodate biotechnological inventions, is examined throughout and forms the subject-matter of the book.

The Patent System

Both in Europe under the EPC and in the US a patent is an intellectual property right. It is, essentially, a negative right granted by the state, which confers on the

---

5   Ibid.

6   See Julian Kinderlerer and Diane Longley, 'Human Genetics the New Panacea?', 1998 vol. 61 *Modern Law Review* 603–20 at 604.

7   See F W A Brom, J M G Vorstenbosch and E Schroten, 'Public Policy and Transgenic Animals: Case-by-case Assessment as a Moral Learning Process', in Peter Wheale, Rene von Schomberg and Peter Glasner (eds), *The Social Management of Genetic Engineering* (1998) Ashgate Publishing UK.

patentee[8] the right to exclude others from using, or exploiting, the invention without his consent. The right is limited in time, under the EPC[9] (and the Agreement on Trade-Related Aspects of Intellectual Property Rights 1994; henceforth TRIPS),[10] to 20 years from the date of filing the application. The right is also limited in terms of space, in that a patent is valid only in the jurisdiction of the patent office by which it is granted. A patent does not confer on the patentee the right to exploit the invention. The right conferred is that of preventing third parties from exploiting the invention without the consent of the patentee. For the most part, exploitation is under the control of national regulatory authorities.[11]

The object of the patent system, both in Europe under the EPC and the US, is to encourage invention and the growth of new industries.[12] The operation of the system is quite simple. The state grants to the inventor rights in his invention in return for public disclosure of how to work the invention.[13] The interests of the inventor are balanced against those of the public. The patent system creates a monopoly in favour of private individuals; hence, there are restrictions. First, a patent can be granted only for an invention.[14] Second, the patentee is given certain exclusive rights in respect of his invention. Third, the exclusive rights in their nature are not permanent.

Specification

In addition, the European patent system, under the EPC and the Directive, contains further restraints in respect of patentability, one of which relates to patent specification. The specification must describe the invention sufficiently clearly to enable those ordinarily skilled in the art to practise the invention. The specification contains a description of the invention and the patentee is granted a temporary monopoly in his invention in return for teaching the public[15] how to work it. The specification, therefore, *discloses* the invention.[16] Additionally, the scope of the monopoly is defined by means of the claims. The claims must be clear, concise and supported by the description.[17] Only if the specification as a whole teaches how to

---

8   The person to whom the patent is granted.

9   Article 63.

10   Article 33. The US has ratified TRIPS.

11   The initiative lies with the patentee, subject to any relevant national regulatory authority, including loss of patent and grant of compulsory licence if the patentee fails to exploit.

12   See generally J M Aubrey, 'A Justification of the Patent System', in Jeremy Phillips (ed.), *Patents in Perspective* (1985) 1–9, ESC Publishing Ltd.

13   However, the manner in which such rights are *used* is controlled by national regulatory authorities.

14   Nothing falling short of an invention can receive patent protection.

15   Or at least those members of the public skilled in the relevant art.

16   Patents Act 1977 s.14(3).

17   Patents Act 1977 s.14(5)(c).

reproduce what the inventor claims can a patent be issued. Adequate disclosure of the invention is the quid pro quo for granting the patentee a temporary monopoly in his invention (Chapter 5).

Criteria

In Europe, both under the EPC and the Directive, and the US, to be eligible for patent protection, an invention must satisfy three criteria:

- novelty
- inventive step (non-obviousness in the US)
- industrial application (utility in the US).

The application of these criteria in practice (Chapter 5) often involves legal subtleties, but they may be simply summarized as follows:

Novelty   This requires that the invention be new.[18] It must not already be available to others, through any form of public disclosure or use before the date of filing the patent application.[19] All such prior knowledge is known as the *state of the art* or *prior art*. Novelty requires a difference from the prior art. Another aspect of novelty is that there must not be *publication* of the invention before the filing date of the application. The rationale for the novelty requirement is that what is known already by the public is not new, and what is already in the public domain can not be the subject of a private monopoly.

Inventive step   This is difficult to define,[20] but is understood as relating to the issue of *obviousness*.[21] The invention must not be obvious to the ordinary skilled worker having regard to the state of the art.[22] In other words, the invention must not merely follow logically from what is already known in the state of the art. Put another way, inventive step requires that the difference in the art possess the character of not being obvious to a person already skilled in that art.[23]

---

18   What is meant by the requirement that an invention be new is discussed in *Merrell Dow v Norton* [1996] RPC 76; see C Floyd, 'Novelty under the Patents Act 1977: The State of the Art after Merrell Dow', 1996 vol. 9 *European Intellectual Property Review* 480–86.

19   Patents Act 1977 s.2(1) and s.2(2).

20   It is not defined in the Patents Act 1977.

21   Patents Act 1977 s.3.

22   The person skilled in the art knows details of the latest developments within his speciality but lacks ability to make inventions – discussed more fully in Ch. 4 hereafter.

23   See generally N J Byrne, *The Scope of Intellectual Property Protection for Plants and Other Life Forms* (1989), report prepared for the Common Law Institute of Intellectual Property. Published by Intellectual Property Publishing Ltd, UK.

Industrial application The invention must be capable of use in any type of industry, including agriculture.[24] The rationale for this requirement is that an invention which cannot be applied in industry is of no benefit to society, and, therefore, is not deserving of patent protection.

Non-patentable Inventions

In addition, in contrast to the US, most European countries have a *list* of subject-matter considered unsuitable for patent protection. Many countries recognize that certain subject-matter is capable of forming the substratum of an invention, yet consider that such inventions ought not to amount to patentable inventions within the meaning of patent legislation.[25] The rationale for exclusion is that, because the inventions are not industrially applicable, they cannot benefit society, and therefore the patentee should not receive monopoly status in respect of exploitation. Indeed, in the context of eligible subject-matter, arguably, it may be stated, as a broad proposition, that in Europe under the EPC and the Directive *immoral* and *biological* inventions are also excluded from patent protection.[26] The rationale is that such subject-matter is not appropriate for patent protection. However, it should be stressed that countries differ in approach to subject-matter patentability, and what is outlined briefly here is the situation which pertains in Europe generally,[27] as a result of both the EPC and the Directive, and in the US.

Because the US patent system and the European system, under the EPC and the Directive, have broadly similar aims and objectives, it might be expected that they would operate in substantially the same manner. However, in this regard, important differences exist and include:

- The origin of intellectual property law in the US is constitutional, meaning that the Constitution of the US confers on Congress the power to enact laws relating to patents. By contrast, in European states the origin of intellectual property law is national and no centralized legislature or enforcement mechanism exists.[28]
- Patents are awarded in the US by the federal government, a centralized US institution. By contrast, in European states under the EPC and the Directive patents are awarded by the European Patent Office, not an EU institution.
- US public policy regards patents primarily as a means of creating new

---

24   Patents Act 1977 s 4(1).

25   Patents Act 1977 s.1(2).

26   See Ch. 2 hereafter for a more accurate and detailed discussion of this statement.

27   As exemplified by the UK in the Patents Act 1977.

28   Although there exists the European Patent Office (under the European Patent Convention) and the EU, which, arguably, can be said to legislate for the whole of Europe centrally on these issues. See generally chapters hereafter.

industries. By contrast, the European Commission regards patents, *inter alia*, as a means of removing barriers to trade, and, hence, promoting the proper functioning of the internal market.[29]

- In the US there is a distinction between patent-grant and exploitation. By contrast, under the EPC and the Directive, if exploitation *would be* contrary to *ordre public* or morality, the patent is not granted.
- The US patent system is older and more tested than is the European system under either the EPC or the Directive.[30]
- In the US the courts developed conditions for patentability. By contrast, in Europe such conditions are laid down by the EPC and the Directive.
- In the US the courts decide what amounts to patentable subject-matter. By contrast, in this regard, the EPC and the Directive are dominant for European states.
- The origin, policy and legislative history of US patent law suggest that a moral regime is not intended. And although, as the cases testify, courts were often willing to withhold patents on inventions they considered immoral, these fell chiefly into only two classes: inventions used to defraud buyers and machines used for gambling (Chapter 3). By contrast, under the EPC, the legislative history suggests that morality embraced wider concerns.[31] No serious attempt is made in the US to deny patents on the basis of moral considerations and, in contrast to European *fora*, courts do not have to deal with such issues as:[32]

  > Whether, and to what extent, moral norms change with time?
  > What are the limits of the morality test?
  > How far into the future can the patent challenger look for the moral effects of the invention?
  > What consensus version of morality applies?

- What is patentable in US law is clear. This is a result of US statutes being drafted in a positive manner, and judicial activism in interpretation. By contrast, under the EPC and the Directive, provisions relating to unpatentable subject-matter are drafted in a complex and negative manner. This can result in patentees not knowing when their actions are safe and lawful.
- US jurisprudence suggests that substantive patent law criteria are conducive to protecting biotechnological inventions, including plant technology.

---

29   Directive on the Legal Protection of Biotechnological Inventions 1998, Recital 5.

30   See Richard Ebbink, 'The Performance of Biotech Patents in the National Courts of Europe', 1995 vol. 75 *Patent World* 25–8 at 26.

31   See generally Ch. 2 hereafter.

32   See Robert Patrick Merges, 'Intellectual Property in Higher Life Forms: The Patent System and Controversial Technologies', 1988 vol. 47 *Maryland Law Review* 1051 at 1062–8; also in Robert Patrick Merges, *Patent Law and Policy-Cases and Materials* 2nd edn (1997), Michie Law Publishers, Va.

Judicial legislation there permits the patenting of life forms and a single regime of protection is possible. By contrast, under the EPC there is a ban on patenting plant and animal variety material, a trend continued under the Directive, resulting in a dual system of protection.

- The existence of the Plant Patent Act 1930 in the US suggests that plant variety material does meet the requirements for patentability.

Biotechnological Inventions

With the advent of biotechnology, the human community stands on the threshold of an extraordinary revolution having profound implications for man and his relationship with other creatures. Therefore, it is not surprising that, despite the patent procedure's basic simplicity, the complexity of modern technology, coupled with the pace at which it is developing, is problematic for patent law. Further problems for inventors have arisen by virtue of the development of both national and international patent laws. The modern inventor[33] must contend with two problems to which the traditional inventor, dealing with mechanical type inventions, was unaccustomed: namely, the nature of the technology itself[34] and the obscurities of international patent laws.[35] The conventional perception of *invention* was conditioned by man's experience in shaping and altering inanimate matter. Manufactured products were built from simpler constituents. Therefore, it was reasonable that any technical application could be characterized by simply listing the structural features. It was also a feature of the classical model that the sequence or steps characterizing the invention were precisely definable.[36]

Biotechnology, dealing as it does with living matter, poses significant challenges for patent law[37] in that biotechnological inventions do not fit as neatly into the classical model as do those of a mechanical nature. Because neither the Patents Act 1977 UK, the EPC nor the Directive is drafted with the special characteristics of biotechnological inventions in mind, difficulties arise when patent law is applied

---

33   Specifically inventors in biotechnology.

34   Biotechnological inventions deal with living matter.

35   International patent law development being somewhat new – discussed more fully in Ch. 2 hereafter.

36   See generally Stephen A Bent, Richard L Schwaab, David G Conlin and Donald D Jeffery, *Intellectual Property Rights in Biotechnology Worldwide* (1987) Stockton Press, NY and Macmillan, London.

37   The claim that a patent is a reward for invention appears untenable in the twenty-first century because of the thin line that exists between invention and discovery in the pharmaceutical, medical and agricultural sectors. See Chika B. Onwuekwe, 'Plant Genetic Resources and the Associated Traditional Knowledge: Does the Distinction between Higher and Lower Life Forms Matter?', in Johanna Gibson (ed.), *Patenting Lives: Life Patents, Culture and Development* (2008) Ashgate Publishing, UK.

to fast-moving emerging technologies such as biotechnology (Chapter 5).[38] The requirements of *novelty* and *inventiveness* are those around which much of the uncertainty arises in deciding whether, and when, biotechnological inventions are patentable. As already noted, the novelty requirement demands that the invention claimed has not been made available anywhere in the world prior to the filing date. Therefore, the question is not whether what is claimed already exists, but whether its existence is known. The inventiveness requirement demands in addition that the invention claimed be not obvious to a person ordinarily skilled in the relevant art. However, it is the author's contention that one of the main problems encountered in patenting biotechnological inventions concerns interpretation of the exclusion provisions.[39]

'Biotechnology' refers to a wide range of techniques that make use of living organisms. The use of biological processes in technology is not new. The making of cheese and wine, or the breeding of plants or animals, are examples of biological processes applied by man for hundreds, if not thousands, of years. However, modern biotechnology is concerned with living organisms and their direct genetic modification. Until recently, the modification of living organisms could occur only by gradual selection. The genes causing the modification of a particular organism could be chosen only from the whole pool of genes of the species to which the organism belonged. Technological breakthroughs in the 1970s offered new possibilities in the manner of effecting gene modification.[40] These techniques built upon the availability of scientific data concerning the structure of DNA.[41] Since all living matter contains the same kind of DNA, it was thought likely that exchange of genetic material between organisms could occur. In 1973 Cohen and Boyer[42] showed that DNA from different species could be assembled and inserted into a host organism. The process of assembling DNA is referred to as *recombinant DNA technology*. Stated alternatively, combining genes of totally different organisms for the purpose of introducing new properties into a host organism is termed *genetic engineering*. The manner in which the process is applied to animals is to modify the genes by insertion of altered DNA, causing them to produce substances normally not produced by them. Organisms whose genetic make-up has been modified in

---

38   See Brad Sherman, 'Patent Law in a Time of Change: Non-obviousness and Biotechnology', 1990 vol. 10 *Oxford Journal of Legal Studies* 278–87.

39   EPC Article 53 – discussed more fully in Ch. 4 hereafter.

40   The three major methods of producing transgenic animals are: microinjection of DNA into the pronucleus, retrovirus infection and *in vitro* establishment of embryonic cells later injected into host cells. See Rainer Moufang, 'Patentability of Genetic Inventions in Animals', 1989 vol. 20 *International Review of Industrial Property and Copyright Law* 823–846.

41   Deoxyribonucleic acid the substance responsible for hereditary characteristics revealed by Watson and Crick in 1953.

42   Stanley Cohen was associate professor of medicine at the University of Stanford, California; and Herbert Boyer was a biochemist and genetic engineer at the University of California at San Francisco.

this manner are termed *transgenic* organisms. Micro-organisms, plants and animals modified in this manner have been used as *bioreactors*[43] for the production of rare pharmaceutical substances. Human insulin and human growth hormone are examples of substances that have been produced by micro-organisms for several years. Today biotechnology promises, inter alia, to feed the world, cure the sick and extend the quality of our lives. Genetic and biological manipulations enable genetic diagnosis, gene therapy, plant-derived vaccines and biopharmaceuticals to tailor health treatment to individual needs.[44]

The main commercial applications of genetic engineering so far have been in the fields of health care, agriculture and the environment. In the domain of health care, biotechnology has been used to produce medicines for the treatment of diseases, such as cystic fibrosis and various cancers. Diagnostic kits have also been produced with the aid of biotechnology. Presently, research is going on into *gene therapy*,[45] which aims to correct congenital disorders. Agricultural biotechnology uses recombinant DNA techniques in animal and plant breeding processes to produce animals and plants with required properties. Presently, transgenic plants include the tomato, potato, sugar beet and tobacco. In environmental biotechnology, micro-organisms have been modified for the purpose of cleaning up soil, water and air.

In today's highly technological world, biotechnology is one of the most research-intensive and innovative industries in the field of science.[46] If European states are to compete in a global environment, arguably, parity of protection with other trading blocs,[47] in particular the US, is necessary. In this respect, strong and harmonized patent rights are required. To this end, the Strasbourg Convention 1963 and the EPC were first adopted. The former was concerned largely with harmonization of substantive patent law criteria, with *optional* exceptions concerning morality. The latter was more concerned with a patent system to support the economic structure of the European Economic Community, with *obligatory* exceptions relating to morality. Under the EPC whether or not patent rights are diluted by moral issues is problematic (Chapter 4). In addition, in this regard, the later adopted Directive is similarly problematic for European Union (henceforth EU) Member States (Chapter 7). In contrast, in the US the Constitution provides for the establishment of patent law without moral restraints. Congress has power to promote the

---

43   Living vessels for the production of pharmaceuticals.

44   See Bita Amani, *State Agency and the Patenting of Life in International Law* (2009) 20, Ashgate Publishing, UK.

45   Gene therapy is the use of nucleic acids as therapeutic medical compounds. The most straightforward gene therapy strategy is to compensate for abnormal gene expression. See Maria A Croyle, 'Gene Therapy', in Daan J A Crommelin, Robert D Sindelar, and Bernd Meibohm (eds), *Pharmaceutical Biotechnology: Fundamentals and Applications*, 3rd edn (2008) Informa Healthcare, USA Inc.

46   See the Economic and Social Research Council report *Intellectual Property Research* (1994) at 18.

47   This is a controversial view. Excessive (patent) monopolies can stifle competition.

progress of science by securing for limited times to inventors exclusive rights in their discoveries.[48] The fact that the US patent system is older and more tested than is the system for European states under the EPC and the Directive suggests it may be a preferred model (Chapter 9). Biotechnology is a vitally important industry both for European states and the US and is a key technology for many industrial sectors including, in particular, agriculture (Chapter 8). Biotechnology promises, and is beginning to deliver, major benefits for European consumers generally and the economies of EU Member States.[49] Although the science of biotechnology has been used for some time, there remains public confusion in European states, and, indeed, the US,[50] over the nature of the technology. In this regard, one of the more persistent, and important, issues of the debate for states involves problems concerning the risks to man and the environment.

Morality Is Now an Issue in Patent Law

The debate concerning the legal, social and moral problems connected with modern biotechnology gives rise to very different attitudes (and interests) not only among the internal participants of the patent system, namely, scientists, economists and lawyers, but also the general public whose concern is mainly twofold: who will determine how the technology is used and who will be the beneficiaries? If the promise of biotechnology is to be realized, wide public debate, incorporating social negotiation[51] that should inform our decisions about the role of law in regulating the development and application of the technology (including the question of generic production), is necessary. However, the real question is whether such control should be exercised in any significant way by means of *moral* considerations in the patent system. The controversy surrounding the role of patents in biotechnology research and product development, and the impact which they can have on those activities, suggest that patent law provides the forum in this biotechnology debate. However, confining the debate to patent law may be misleading. On the one hand, the question of morality in essence concerns the act of creating the technology and as such is problematic within the patent system, since patent law is concerned with protection of the technology once it exists. Arguably, however, on the other hand, opposing a patent on moral grounds is tantamount to preventing the activity

---

48   Constitution, Art. 1, s.8, clause 8, discussed more fully hereafter in Ch. 3.

49   In the period between 1995 and 1996 the number of European biotechnology companies grew by 23 per cent to over 700, employing some 27,500 people; see Graeme Laurie, 'Biotechnology: Facing the Problems of Patent Law', *Innovation, Incentive and Reward* Hume Papers on Public Policy vol. 5 (1997) 46–63 at 47, Edinburgh University Press.

50   Less so in the United States.

51   Discussed more fully in Ch. 7 hereafter, where concerns of the European Parliament in respect of the Biotechnology Directive 1998 are noted.

altogether, by a withdrawal of the incentives to perform it, and suggests that patent law is a component in regulating, albeit indirectly, the creation of biotechnology. However, there seems little evidence to suggest that morality provisions in patent law are there to regulate per se. In the first place, morality provisions have no de jure jurisdictional basis and, secondly, it is not clear from the cases that a regulatory approach de facto exists (Chapter 4). Where moral considerations most apply is in elucidating concerns to determine what is and what is not acceptable to society and how these should be dealt with. That in itself is a legitimate reason for pursuing the moral assessment of biotechnology. While scientists are capable of identifying the benefits of biotechnology, they are no better equipped than anyone else to address the social and moral issues that might arise as a result of using the products thereof. Only the general public can ultimately decide the degree of importance to be attached to the benefits, the hazards and their impact. Patent law has traditionally steered clear of moral arguments (Chapter 4). However, moral arguments enter the patent arena directly through the gateway of Article 53 of the EPC. Essentially dormant until the advent of biotechnology, Article 53 provides, *inter alia*, that patents shall not be granted for inventions the exploitation of which *would be* contrary to *ordre public* or morality. Through this gateway have come arguments which, previously, were not considered pressing issues in patent law. Biotechnology has changed all that. Patent law is now one of the central areas in which moral issues are raised.

Moral Objections to Patenting

It is the author's contention that patent law is not designed, or, indeed, appropriate to regulate biotechnology and any attempt to do so, in particular by denying patents on the basis of morality, is misplaced as such a solution does not match the nature of the problem. One reason is that the predominant aim of patent law is to encourage investment and innovation through strong protection. There is little doubt that economic policy lies at the heart of, and has been advanced by means of, patent law.[52] In contrast, the aims of a moral policy (if one exists at all) are unclear (Chapter 4) and can not be achieved, or achieved adequately, through the patent system.[53] The issue here is whether or not a policy designed to advance economic interests can adequately accommodate moral concerns.[54] A policy enshrining moral values suggests taking a stance on patentability with clear consequences for Europe's economic future. Accordingly, it is imperative to ascertain the aims of a moral policy. Once these are established, it then becomes necessary to determine whether or not the economic and moral policies are of equal weighting. If yes, the

---

52   See Graeme Laurie, 'Biotechnology: Facing the Problems of Patent Law', *supra* at 61.

53   Ibid. at 61.

54   Ibid. at 61.

question then arises: can the underlying aims of each policy be achieved by their incorporation into patent law?

On the one hand, conceptualization of the issues by the participants, namely, those of science, economics and law, is that a strong patent system is a core aspect of commercial development. Economic arguments are always to the fore when justifying the patent system.[55] Simply put, the basis of the main economic argument is that it is just to reward the inventor for his time, skill and effort in creating the invention. The inventor must be permitted the possibility of recouping his investment without being forced by competitors to lower his price, so that he might never be in a position to earn back invested money. According to this argument, exclusivity is justified on the basis that it is a necessary element of incentive for research and development. Another argument put forward when justifying the patent system is that patents help to stimulate innovation and industrial development through dissemination of technical knowledge. The patent system, operating as it does on a disclosure basis, ensures that details of the invention are made available to the public. Researchers are then free to build upon this knowledge and thereby further advance the state of the relevant technology. This argument also works in reverse. If a technology is excluded from patentability, there is no incentive to invest in research. In such a situation the public may be deprived of knowledge and of any advantage the technology might have to offer.

On the other hand, however, not everyone accepts the presumption that patents are good, and the extent to which research and development would suffer as a result of withdrawing the incentive is unclear. In the first place, patents do not protect each inventor who conceives the invention.[56] It is the first to apply for a patent, rather than the first to invent, who is given priority.[57] Secondly, since inventions are there to be developed, industries that have progressed to a certain point will inevitably make them, and so any artificially created incentive is unnecessary.[58] Thirdly, in certain circumstances countries now require a lengthy testing of new products, meaning the period of protection on the market is reduced.[59] For these reasons doubt must surround the role of patent systems in encouraging the exploitation of major inventions. This applies, in particular, to biotechnological inventions. In addition, the granting of patents is opposed by many *outside* the

---

55   Other justifications include: the 'natural law' thesis, the 'incentive thesis' and the 'contract' thesis.

56   See W R Cornish, *Intellectual Property: Patents, Copyright, Trade Marks and Allied Rights* 4th edn (1989) 79, Sweet & Maxwell.

57   The US operates a 'first to invent' system; the Patent Reform Act 2007, however, introduces a 'first to file' system, similar to EPC requirements.

58   W R Cornish, *Intellectual Property: Patents, Copyright, Trade Marks and Allied Rights* 2nd edn (1989) 80, Sweet & Maxwell.

59   In the pharmaceutical industry, clinical trials lasting 11 to 12 years are not uncommon before the product is allowed on the market, meaning the period of protection is accordingly reduced.

system[60] on grounds more fundamental than traditional arguments and this is the reason why morality is now an issue in patent law despite not being particularly so in the past. Opponents[61] of genetic engineering inventions argue that granting monopoly rights to the patentee to exploit such inventions is simply inappropriate. The argument is premised on the belief that intellectual property rights cannot, and must not, be placed above the right of all human beings to live a full and productive life. In this regard, the right to development is paramount. The patent system challenges the realization of the right to 'development', particularly in the context of the right to food and health. Opponents of the patent system also argue that the high level of protection afforded by existing national and international regimes tilts the somewhat already delicate private/public balance in favour of private interests.[62] Given the recent groundswell of public opinion in this regard, and the revised Article 52 and Article 53 EPC, such an argument deserves careful consideration (Chapters 4 and 6).

Patents for biotechnology are being objected to also on the grounds, not that only *one* person should be able to conduct the activity for which the patent is sought, but that *no one* should. The main criticism is that to allow patenting of new human or animal traits is to condone the commercialization of life, and this is morally insupportable. The Humane Society of America has termed genetic engineering 'a step backwards in the evolving recognition of the significance of animal life, the sanctity of being and the interconnectedness of all life'.[63] The objection is rooted in concepts of property and ownership, and the appropriateness of the application of these concepts to the DNA of *humans* in particular (no one seriously disputes that a dairy farmer owns his cows and can sell or slaughter them at will, therefore 'owning life'; see Chapter 7). The argument rests on the 'slippery slope' premise; that is, once patents are granted over animals, it is only a matter of time before they are granted over humans. However, such an argument presumes that it is impossible to draw a principled distinction between animal and human life. It also presupposes that patenting of animal life will, as a matter of fact, cause human life to be seen as a commodity. The fear is that such a shift in thinking will ultimately be responsible for allowing patents on humans. The argument has a couple of limbs. One is that allowing patents on animals will cause a change in attitude towards human life. The other is that any change occurring will assimilate human life to animal life. The reasoning presupposes that any change will involve a 'levelling down' process for humans, whereas it could be viewed as a 'levelling

---

60   For example such organizations as Greenpeace and Friends of the Earth as well as religious groups.

61   Ibid.

62   See Adojoke Oyewunmi, 'The Right to Development, African Countries and the Patenting of Living Organisms: A Human Rights Dilemma', in Johanna Gibson (ed.), *Patenting Lives: Life Patents, Culture and Development* (2008) Ashgate Publishing, UK.

63   See Michelle Paver, 'All Animals Are Patentable, But Some Are More Patentable Than Others', 1992 vol. 9 *Patent World* 9–15 at 14.

up' process for animals. In response, while 'commercialization of life' concerns need to be addressed, the issue is one of regulation involving specialized agencies as appropriate.

However, it must be accepted that, on balance, patent law is primarily an instrument of economic policy. It provides an incentive to invest and innovate (the issue of whether or not the patent system encourages firms to engage in unnecessary duplication of effort has provided a lively debate among economists). Further, it provides a centralized information-gathering system so that other inventors can gather ideas or develop new inventions. On this basis, those in favour of patenting argue that an incentive mechanism is essential to maintain the creation of the benefits flowing from biotechnological research including, in particular, benefits for health care, agriculture and protection of the natural environment.

Intrinsic Objections to Biotechnology per se

Arguments against biotechnology per se suggest that it is the creation of such inventions which is problematic. The core objection seems to be that biotechnology, and more specifically genetic engineering, is wrong in itself, even if the net benefits outweigh the harm caused. Genetic engineering is considered by many to be intrinsically wrong for the following reasons:[64]

- It is an attempt at 'playing God'.
- Genes represent the common heritage of mankind and should be passed from generation to generation without technical intervention by man.
- Genes occur naturally in organisms and should not suffer interference.

Basically, the argument is that, no matter how good humans are at biotechnology, they simply should not attempt it.[65] On the one hand, this approach is simply 'adopting a position' rather than justifying it. Do not 'birth control' techniques control the sexual process to satisfy desire? Where is the moral outcry here? In short, any interference in the process of nature can be described as 'playing God'.[66] On this basis, the mere fact that genetic engineering allows control over *life* is not a

---

64   See Sigrid Sterckx, 'European Patent Law and Biotechnological Inventions', in Sigrid Sterckx (ed.), *Biotechnology, Patents and Morality* (1997) 1–55 at 4, Ashgate Publishing, UK.

65   In 1980 the Human Genome Project was first proposed. Its purpose was to sequence and define all of the human genes. The objectives were: sequence the genomes of humans and selected model organisms; identify all of the genes; develop technologies required to achieve all of the above. The sequencing was completed in 2003. See Nathan S Mosier and Michael R Ladisch, 'Genomes and Genomics', in *Modern Biotechnology: Connecting Innovations in Microbiology and Biochemistry to Engineering Fundamentals* (2009), Wiley & Sons.

66   See Barry Hoffmaster, 'The Ethics of Patenting Higher Life Forms', 1988 vol. 4 *Intellectual Property Journal* 1–24 at 4.

sustainable moral objection. Unless, and until, control which humans express over nature by virtue of genetic engineering can be differentiated from control which humans routinely exercise over nature in other ways, then genetic engineering must remain on a moral par with any other human-made intrusion into nature.[67] On the other hand, there is a moral (as well as a legal) obligation on an individual to be accountable for benefits conferred upon him by the government. In this regard, far from the patent system not being the appropriate place to consider moral issues, it has often been used explicitly to exclude certain inventions from patentability due to 'moral' considerations. Patent regimes have continually acted as a moral filter, allowing certain forms of culture to pass into mainstream commercial life and blocking others.[68] Restrictions on patentability have permitted courts and patent assessors to make value judgments on issues of social advantage. An example is the patenting of nuclear weapons technology prohibited in the US.[69]

Could the fact that genetic engineering is a risky business, whose consequences as yet remain largely unknown, be a legitimate moral argument against biotechnology?[70] (See Chapter 4.) The answer to this question is likely to be 'no' because, in this respect, it is the same as any other new technology. The most profound positive, and negative, results remain unknown at the early stages of any technology. Safety concerns are not unique to the field of genetic engineering. Since safety risks are specific, each must be assessed in its own context.

Another moral argument against biotechnology is that animal testing for genetic engineering purposes is wrong because pain and suffering are inflicted upon animals for ends that appear frivolous in contrast.[71] It is difficult to sustain such an argument because it is based upon an absolutist approach enshrining one value only, namely, that of protecting animals. Moral decision-making involves a continuous accommodation of conflicting values.[72]

To be credible, any moral argument must offer an explanation about what is so different, and unacceptable, about genetic engineering. Such an explanation is made more difficult by virtue of the fact that (like the US Patent and Trademark Office), by permitting patents on life forms, the European Patent Office has apparently 'married' science and profit. On a purely moral basis, should science not be research-driven?[73]

---

67   Ibid. at 6.

68   See Angus J Wells, 'Patenting New Life Forms: An Ecological Perspective', 1994 vol. 3 *European Intellectual Property Review* 111–18 at 112.

69   42 USC s.218(a) (1982).

70   See Barry Hoffmaster, 'The Ethics of Patenting Higher Life Forms', *supra* at 6–7.

71   Ibid. at 8.

72   Ibid. at 9. See also Richard Ford, 'The Morality of Biotech Patents: Differing Legal Obligations in Europe?', 1997 vol. 6 *European Intellectual Property Review* 315–18 at 318.

73   Although, if not profit-driven, science may be less likely to achieve.

It is difficult to characterize what exactly it is about genetic engineering that does not allow it, in the opinion of objectors, to go *wrong* compared with other technologies. Can a higher standard of morality be justified on the basis that such technology deals with *life*, or that the consequences of it going wrong would be catastrophic? Since the former concern is not unique to biotechnology, the imposition of a higher standard of morality is not easily justified. And the latter concern is one of safety, which can adequately be dealt with by proper regulation as with any other technology.

Any science, including genetic engineering, will morally amount to a 'mixed blessing', having both advantages and disadvantages. The question is: where is the line to be drawn? What will society accept in terms of a tolerably *bad*, or undesirable, outcome of genetic engineering? It is the author's contention that in the context of patent law the function of morality is to elucidate concerns to determine what is and what is not acceptable to society. Thereafter, specialized agencies can, by proper regulation, be controlling.

In regard to micro-organisms, the concern is that modified organisms can escape into the environment from the laboratory in which they are created, without scientists knowing the effect this will have for future generations. In regard to plants, the concern is essentially the same as it is for genetically modified micro-organisms, namely, that modified plants should not be released into the environment, since the ecological balance may be disturbed.[74] Opponents[75] of genetic engineering are not appeased by the argument that, once it is established that the modified plant is harmless, there can be no ill-effects on the environment. And, in the case of plants, there is an additional objection, namely, that genetic modification will cause a reduction in biodiversity. 'Biological diversity' means the variability among living organisms from all sources, including, *inter alia*, terrestrial, marine and other aquatic ecosystems and the ecological complexes of which they are part; this includes diversity within species, between species and of ecosystems.[76] On the one hand, the argument is that, in the long term, only those plants serving the interests of man will remain and biodiversity will (ultimately) be unacceptably reduced. On the other hand, it may be that biodiversity will be enriched by the introduction of non-plant (or cross-plant) genes into existing plants.

---

74 While plant biotechnology has been very successful in transforming plant-agriculture, there are still concerns about safety and sustainability. Such innovations which will find their way into future generations can be controlled by means of precision engineering. This technique can also alleviate some bio-safety concerns. See C Neal Stewart Jr. (ed.), *Plant Biotechnology and Genetics*, ch. 16 (2008) Wiley & Sons, USA.

75 Such as Greenpeace, Friends of the Earth and religious groups.

76 See the United Nations Convention on Biological Diversity, Art. 2, [1992] vol. 31 ILM. The place of patents in meeting the objective of the Convention is discussed by Charles Lawson, 'Patents and Biological Diversity Convention, Destruction and Decline', 2006 vol. 28 *European Intellectual Property Review* at 418.

In regard to animals, the arguments against genetic modification run parallel to those for plants. Danger of disruption to the ecosystem, caused by release of transgenic animals into the environment, is the primary concern. The biodiversity reduction argument also applies for animals according to opponents, who maintain that only those animals beneficial to man will remain. On the one hand, the claim is reinforced by the observation that properties of animals are changed in a way that is beneficial to man, but only rarely for the animals themselves. On the other hand, biodiversity may be enriched as a result of non-animal (or trans-animal) genes being introduced into existing animals.[77]

An additional objection is one concerning experimentation with animals (Chapter 4). A host of questions come to mind. Are animal experiments necessary? Is it permissible to conduct animal experiments for purposes other than the health and well-being of the animals themselves?[78] Should the use of animals as bioreactors be prohibited in all cases, or should it be permitted in certain circumstances, such as the production of pharmaceuticals? Whatever the merits, or demerits, of these arguments, the relevant question is[79] whether or not the harm to animals is outweighed by the benefit to humans.[80] The philosophical issue is whether or not genetic manipulation violates animal *integrity*.[81] In addition, concern has been raised over negative impacts which transgenic animals might have on their own species. The concern is based on the assertion that transferring genes between species transgresses natural barriers between them, thus violating species integrity.[82]

The most complex issues arise when considering genetic modification of human beings (Chapters 7 and 9). Viewed from the public perspective the threat posed by contemporary biotechnology is the possibility that it will alter human nature in an irrevocable manner. Some of the more important questions arising are:

- Do we, as humans, *own* our genetic material, or does it belong to society as a whole? Does the 'common heritage' argument, i.e., that material

---

77   Arguments for, and against, patenting transgenic animals can be found in *New Developments in Biotechnology*, Office of Technology Assessment Special Report no. 5, US Congress (April 1989), Washington, DC.

78   For example the fabrication of cosmetics.

79   According to George Annas, 'the real question is … [n]ot whether animals can or should be patented, but what things it is reasonable to permit humans to do to animals'. See generally George Annas, *Of Monkeys, Man and Oysters* (1987) Hastings Center Report vol. 17.

80   Discussed more fully in Ch. 4 hereafter.

81   See generally *New Developments in Biotechnology*, Office of Technology Assessment Special Report no. 5, *supra*.

82   Ibid.

possessed in abundance by vast numbers of people cannot be the subject of a private monopoly, apply.[83]

- Is intervention into the human genome an attack on human dignity?[84]
- Is gene therapy, i.e., the provision of healthy copies of flawed genes, acceptable? Proponents of genetic engineering argue that intervention into the human genome is necessary, ultimately leading to an increase in human biodiversity; while opponents claim such a step is turning the sacred into the profane.

Whether public opinion is as a result of well-informed and reasoned thought, or merely instinct, is unclear and difficult to determine. It is the author's contention that elucidating the moral concerns associated with biotechnology could clarify the fears of the general public and address how these should be dealt with by society.

Conclusion

Need patent law's response to new technology, in particular biotechnology, necessarily be informed by moral considerations? To what extent should moral considerations apply and, if they do, how are they to be dealt with by society? Traditionally, patent legislation essentially regulates economic and competition issues and not wider considerations outside the commercial area. However, in contrast to the United States, at present this can not be the case for EPC or EU Member States because they are subject to a variety of legal instruments each containing morality-based constraints in respect of patentability. The result is that confusion as to the obligations of states exists and is likely to continue.

---

83   See Barbara Looney, 'Should Genes Be Patented? The Gene Patenting Controversy: Legal, Ethical and Policy Foundations of an International Agreement', 1994 vol. 26 *Law and Policy in International Business* 231–72 at 234 – discussed more fully in Ch. 7 hereafter.
84   Ibid. at 239.

Chapter 2
Development of a European and UK Moral Regime

Introduction

The patenting process stands at the confluence of science and technology, on the one hand, and law, on the other.[1] The intertwining of two disciplines has in the past created certain tensions even at the national level, evidenced by the fact that certain sectors of industry have received less patent protection when compared to others. This was sometimes the case, for example, for foodstuffs and pharmaceuticals where patent protection was deemed inappropriate.[2] These tensions became more pronounced as companies moved and sought patents outside their country of origin. While the scientific and technical side remained the same the world over, in particular in Europe and the United States, divergence in national patent laws meant an almost endless variety of rules on conditions, procedure and scope of protection. Because the point of view of the inventor is essentially technical, he reasons from the basis of the invention made, for which he claims full protection. However, the view of the national legislator is necessarily more complex and is influenced by several factors, including:

- different concepts justifying patent protection;[3]
- patent law as a matter of economic policy;[4]
- patent law as an integral part of the general legal system of the country concerned.[5]

---

1   See generally M van Empel, *The Granting of European Patents* (1974) ESC Publishing, UK.

2   Discussed more fully hereafter.

3   Traditionally four concepts are distinguished: the 'natural law' or 'intellectual property' thesis, the 'reward' thesis, the 'incentive' thesis and the 'contract' or 'disclosure' thesis.

4   Of the four concepts listed here, those which stress the function of patents as an instrument of economic policy seem dominant at present.

5   Enforcement of patent rights is tributary to the general rules on civil procedure. Granting procedure, too, will normally be subject to administrative control and the administration of justice, as these prevail in the country concerned.

Against this background it can readily be appreciated why progress, in particular for European states, in the 'internationalization' of patent law was slow initially. However, as the number of patent applications filed abroad grew, the call from industry for uniformity and centralization grew more intense. The response of states was the introduction of a number of legal instruments, in particular the Strasbourg Convention 1963 and the European Patent Convention 1973 (hereafter EPC), each designed to ensure the progressive harmonization of patent law.

There was no such thing as European patent law prior to 1973. The Strasbourg Convention 1963 did exist, but this was predominantly concerned with reshaping substantive provisions of patent law and did not deal in any meaningful way with moral concerns. However, after 1973 patent law became subject to the provisions of the EPC. This was largely based on the Strasbourg model. In this chapter, the development of the EPC is examined against the background and aims of the Strasbourg Convention. In particular, the provisions of Strasbourg which allowed states the *option* of precluding patentability on moral, and other, grounds are assessed, and reveal the intention was to continue practice under the corresponding provisions of national laws, a trend followed in the EPC with one important difference, namely, the latter provides for *mandatory* exclusions from patentability. Also examined in this chapter in terms of morality is national patent practice in the UK and shows that, historically, a 'light' regime, largely concerned with personal morality, was adhered to. The idea of attaining a particular moral standard to be presented as a general criterion of patentability was never considered. By contrast, the legislative history of patent law in the United States and the approach to the question of morality is outlined and discussed in Chapter 3, while the legislative history of the later adopted Biotechnology Directive 1998 is outlined and discussed in Chapters 7 and 8.

International Patent Law Development

The development of international patent law was a response to the needs of industry for greater, and more uniform, protection around the world. To meet these needs, legal instruments, some of which are outlined briefly hereunder, were adopted, and institutions established at the international, regional and national levels.

Paris Convention for the Protection of Industrial Property 1883[6]

The Paris Convention laid the foundation for all further unification in the area of industrial property, hence its importance. In the first place, it did away with discrimination and the reciprocity requirement between states by providing for

---

6   Subsequently revised in 1900 (Brussels), 1911 (Washington), 1925 (The Hague), 1934 (London), 1958 (Lisbon), 1967 (Stockholm). The UK ratified the Stockholm Act in 1969.

'national treatment'. Under Article 2 of the Convention, each Member State must apply to nationals of other Member States the same treatment as for its own nationals in respect of patentability criteria. In the second place, the Convention established a common minimum standard as to the protection of industrial property. In this respect the establishment of a 'priority' regime was of great practical importance. By virtue of Article 4 of the Convention, an applicant who has duly filed an application for a patent, utility model, industrial design or trade mark, in one of the countries of the Union, enjoys a right of priority for the purpose of filing in other Member States. This means that an application in one of the states gives a period of time in which to pursue an application in any of the others, which will bear the same priority date (date of filing the original application) as the first. In this manner, the applicant is able to secure a 'filing date' (a date from which the patent will run) and after which other disclosures do not form part of the prior art by which to test the novelty of the application. The Brussels Revision of 1900 set this period at 12 months. Due to the large number of states which adopted the Convention, it was difficult to achieve further development within the actual framework of the Paris Union itself. However, the Convention provided for such a situation and introduced an element of flexibility. Article 19 of the Convention reserves the right for countries of the Union to make separately between themselves special agreements for the protection of industrial property, in so far as these agreements do not contravene the provisions of the Convention. The use of 'special agreements',[7] rather than adapting or extending the 'mother' Convention, thus facilitated 'internationalization' of patent law.

Patent Cooperation Treaty 1970

That 'search' and 'grant' procedures should be separated was recognized when the director of the Bureaux internationaux réunis pour la protection de la propriété intellectuelle (BIRPI)[8] was invited in 1966 to undertake a study on solutions to reduce duplication of effort by applicants and national patent offices. The study finally resulted in the Patent Cooperation Treaty 1970 (hereafter PCT),[9] the main features of which can be summarized as follows.[10]

---

7 The European Patent Convention in its Preamble is expressed to be a 'special agreement' within the meaning of Article 19 of the Paris Convention.

8 The International Office for the Protection of Intellectual Property. This was the international authority that administered both the Paris Convention and the Berne Convention (1886) for the Protection of Literary and Artistic Works. Under the 1967 (Stockholm) Act Article 15 of the Paris Convention BIRPI is now replaced by the International Bureau of Intellectual Property, the Secretariat of the World Intellectual Property Organization (WIPO).

9 Concluded at Washington, DC on 19 June 1970.

10 See generally M van Empel, *The Granting of European Patents*, *supra*.

An applicant may institute applications in numerous countries by a single procedure and may delay a final decision to apply in a number of countries for a period of 20 months. As well as providing for an 'International Search Report', the PCT contains a Chapter dealing with the so-called 'International Preliminary Examination'.[11] The objective of the preliminary examination is to formulate a preliminary non-binding opinion on the question of whether or not the claimed invention appears to be novel, involves an inventive step and is industrially applicable.[12] The PCT sets out the criteria by which each of these requirements is to be measured. However, because criteria of patentability differ from one country to another, any cooperation in the field of examination of applications can lead only to a non-binding opinion.

The Strasbourg Convention 1963

The Strasbourg Convention 1963 was the basis for the EPC. The development of the European patent can largely be attributed to the work of the Council of Europe. The Consultative Assembly of the Council of Europe in Recommendation Number 23 of 8 September 1949 proposed the drafting of a convention on the creation of a European patent office. Following this recommendation, a Committee of Experts was set up[13] to examine a multitude of plans submitted to the Consultative Assembly of the Council.[14] The ultimate long-term aim was to establish a European patent. However, by the mid-1950s it was generally accepted that the necessary first step towards a European patent was harmonization of national laws and procedures. The Formalities[15] and Classification[16] Conventions were the first steps to this end. However, the real challenge lay in a harmonization of substantive patent law.

As already noted, in 1955 a 'Committee of Experts on Patents' was set up under the aegis of the Council of Europe to study the basic components of patent law. Issues to be examined included:

- novelty;
- technical progress and creative effort;
- scope of description and claims.

---

11   PCT Ch. 2. The PCT sought a move from 'registration' to 'examining' systems for states.

12   PCT Article 33.

13   In 1955 the Committee of Experts was established by the Council of Europe and Reports were made and debated in 1961. See Document 1708 vol. 75 session 15, 1963/64.

14   See generally M van Empel, *The Granting of European Patents, supra.*

15   European Convention Relating to the Formalities Required for Patent Applications 1953.

16   The Council of Europe Convention on the International Classification of Patents for Invention 1954.

Reports on each of these questions were made in 1961 and coordinated by the head of the French Patent Office.[17] By 1962 the reports had given rise to the main features of a convention on substantive patent law. In the following year the Strasbourg Convention was adopted.[18] The Strasbourg Convention required states to compromise considerably on, and to rethink fundamentally, the concepts underlying existing patent law. Examination of some of the more important questions reveals, first, how intellectual property concepts can be reshaped when the political will exists to do so, and second, what were seen as the essential elements of the proposed new model for substantive law. In regard to morality, that a 'permissive exclusion' was agreed among Contracting States suggests that such restraints were not essential elements of the patent system but on the margins.

The Strasbourg process of harmonization presented European states with the opportunity to reshape traditional patent law concepts. The arrival of new technology, in particular biotechnology, offers a similar challenge now. However, international agreement on substantive patentability criteria may not in itself be sufficient to result in harmonization (Chapter 5). A consensus on the part of states as to the place of morality within the patent system is also required. Whether or not such a consensus exists is difficult to determine and European Patent Office jurisprudence suggests that the Office is unable to offer guidance for future cases as to the meaning and application of the morality provision (Chapter 4).

At Strasbourg the reshaping of essential concepts of patent law from the UK perspective included:

Novelty At this time[19] the Patents Act 1949 was in force in the UK. The definition of prior art against which novelty was judged excluded patent specifications more than 50 years old,[20] and anything not known in the UK.[21] The novelty test was, therefore, 'local and limited'. If harmonization were to become a reality, the possibility of each country having a local novelty test would have to be ruled out. The Convention adopted the 'absolute' novelty approach whereby the prior art against which novelty was to be judged was universal.[22] The harmonization achieved at Strasbourg resulted in an international standard being adopted for the novelty criterion and was to inform future patent instruments.[23]

---

17   Mr Finniss was the then President of the French Patent Office.

18   The Convention on the Unification of Certain Points of Substantive Law on Patents for Inventions, Strasbourg 27 Nov 1963, Council of Europe. *European Treaty Series* no. 47 (Cmnd 2363 HMSO 1964).

19   The 1960s.

20   Patents Act 1949 s.32(1)(e).

21   Patents Act 1949 s.50(1).

22   Strasbourg Convention 1963 Art. 4(2).

23   Patent Cooperation Treaty 1970 and the European Patent Convention 1973.

Inventive step   It was common ground among European states that a patentable invention should be distinguished from the prior art by something more than mere novelty. However, national laws varied considerably on this matter. German law, for instance, required the presence of an 'inventive step'. However, German jurisprudence qualified this requirement by introducing doctrines of 'inventive height' and 'technical progress'. UK law spoke of an invalidity test where the invention did not involve an inventive step and was obvious.[24] Although it was difficult to define the term 'inventive step', the concept was associated with non-obviousness,[25] and Germany supported the UK in committee by agreeing to omit any reference to technical progress. Therefore, in the final draft, Article 5[26] of the Convention spoke of invention thus: 'An invention shall be considered as involving an inventive step if it is not obvious having regard to the state of the art.' The Strasbourg Convention thus laid down a formula for inventive step which was to prove acceptable to future patent harmonization instruments.[27]

Industrial application   Another major element of substantive law discussed by the Strasbourg committee was the definition of 'patentable invention'. It was, again, common ground that inventions in manufacturing industry were patentable. However, in this regard it should be observed that at the time many countries excluded pharmaceuticals[28] and foodstuffs[29] from patentability.[30] The reason for this was a belief that the protection of life and health are universally recognized objectives and transcend proprietary rights. Additionally, in various countries there were limitations on the patentability of living processes.[31] The result was that protection was not available to all technical fields. In the UK in particular, agricultural and horticultural processes were regarded as unpatentable simply on the basis of judicial interpretation of what was meant by 'manner of new manufacture' within section 6 of the Statute of Monopolies of James 1.[32] The Statute of Monopolies banned all monopolies except those saved by section 6, which reads as follows:

> [a]ny letters patent and grant of privilege for the term of 14 years or under
> hereafter to be made of the sole working of any manner of new manufacture

---

24   Patents Act 1949 s.32(1)(f).

25   This was true, too, for other countries whose legal systems were British-based, as well as for the US.

26   First sentence in text.

27   Patent Cooperation Treaty 1970 and the European Patent Convention 1973.

28   Such as Germany and Italy.

29   Stimulants and medicaments as well as substances manufactured in a chemical manner were also excluded from patentability.

30   See generally Rainer Moufang, 'Methods of Medical Treatment under Patent Law', 1993 vol. 24 *International Review of Industrial Property and Copyright Law* 18–49.

31   Mainly for historical reasons.

32   1624, passed in the twenty-first year in the reign of James 1.

within this realm to the true and first inventor and inventors of such manufactures … [s]o as also they be not contrary to law or mischievous to the State.

Section 6 was incorporated into section 101 of the Patents Act 1949 and read as follows:

'Invention' means any manner of new manufacture the subject of letters patent and grant of privilege within section six of the Statute of Monopolies and any new method or process of testing applicable to the improvement or control of manufacture, and includes an alleged invention.

However, around the time of the Strasbourg negotiations, the position in the UK as regards what the Act meant by 'new manner of manufacture' was broadened by judicial decision to include agricultural and horticultural processes.[33] This allowed UK courts to interpret 'industrial character' in as wide a manner as possible, thus extending patent protection to all technical fields.

Scope of Description and Claims of the Strasbourg Convention

It was also common ground at the Strasbourg Convention that the claims defined protection for the invention. However, the issue arose as to whether the scope of protection should be limited by the words used in the claims,[34] or whether protection should extend to the general inventive concept inherent in the claims,[35] or whether it should lie somewhere between these two extremes. The question also arose as to whether the claims could be taken to cover something described in the specification but which, as a matter of language, lay outside the claims. In a compromise on these issues, it was agreed[36] that the description could be used to interpret the claims[37] and that 'The extent of protection conferred by the patent shall be determined by the terms of the claims'.[38] In this regard, the compromise reached at Strasbourg laid the foundation for the subsequent adoption of the 'Protocol on the Interpretation of Article 69 of the European Patent Convention'.[39] At Strasbourg, therefore, there emerged a patent system *essentially* universal in character and capable of extension to all technical fields. The Convention established that substantive criteria for patentability enshrined concepts of novelty, inventive step and industrial character, which could be reshaped to meet new needs.

---

33   See *Swift & Co's Application* [1961] RPC 129.
34   The approach adopted in the UK.
35   The approach adopted in Germany.
36   It was accepted that 'terms' was looser 'words' but tighter than 'inventive concept', the approach previously adopted by the Germans.
37   But not to add to them.
38   Strasbourg Convention 1963 Art. 8(3).
39   Patents Act 1977 s.125.

Despite the universal character of the Strasbourg Convention, some exceptions to patentability had to be admitted for politico-economic reasons. In the context of morality, two exceptions were permitted, but were *not mandatory*, on a permanent basis, and are outlined in Article 2 of the Convention.

Morality provision under Strasbourg   Moral criteria informing validity were never a central concern of Strasbourg, although they were mentioned in the final text. The first exception to patentability *permitted* in the Strasbourg Convention concerned the refusal of patents for inventions, the publication or exploitation of which *would be* contrary to *ordre public* or morality. The exception was outlined in Article 2(a) in the following form:

> The Contracting States shall not be bound to provide for the grant of patents in respect of:
>
> (a) Inventions the publication or exploitation of which would be contrary '*ordre public*' or morality, provided that the exploitation shall not be deemed to be so contrary merely because it is prohibited by law or regulation.

Armitage and Davies,[40] who were both closely involved in the preparatory work of the Strasbourg Convention, state that the morality provision did not feature in the early drafts of the Convention.[41] As noted hereafter, historical documentation relating to the creation of Article 53 of the EPC, which is based on Article 2 of the Strasbourg model, supports Armitage and Davies, who further contend that morality was not meant to be concerned with the essentials of patent law but was added to permit the continuation of powers existing in national laws to refuse patents where the granting of them would be unacceptable on moral or public order grounds. The morality provision of Strasbourg, therefore, was intended to recognize two legitimate government concerns. One was that governments should not have to publish obscene documents, and the other was that they should not have to publish instructions on how to perform acts leading to a breach of the peace or breakdown of morals.

Variety provision under Strasbourg   The second exception to patentability *permitted* under Strasbourg is also a concern about morality in that it relates to not merely plant varieties but also animal varieties (Chapter 4). In this regard, the second exception was very much a product of its time. The feeling of the day among states was that plant varieties were best protected by subject-specific

---

40   Edward Armitage and Ivor Davies were both former Comptrollers of the UK Patent Office.

41   See generally Edward Armitage and Ivor Davies, *Patents and Morality in Perspective* (1985) ESC Publishing, UK.

national legislation.[42] Alternatively, plant varieties could receive protection by virtue of the International Convention for the Protection of New Varieties of Plants 1961 (henceforth UPOV 1961). At this time, too, there was a strong belief that there should not be double protection of any category of entities.[43] Hence the ban on double protection in the 1961 text of the UPOV Convention.[44] Additionally, although drafted with plant varieties in mind, the extension of the exception to animal 'varieties' was a logical consequence for two reasons. First, it would have been anomalous to refuse protection to a certain category of plants, while awarding it to the corresponding category of a higher *life form*. Second, it was thought prudent to make allowance for the possible creation in the future of a system of protection for animal breeds along the lines of the known plant variety protection (this has not yet occurred).

The exclusion of patent protection for plant varieties under UPOV 1961 was an obligation for UPOV Member States only. Consequently, it was provided in the Strasbourg Convention[45] as *optional*, in the following form:

> The Contracting States shall not be bound to provide for the grant of patents in respect of:
>
> (b) plant or animal varieties or essentially biological processes for the production of plants or animals; this provision does not apply to micro-biological processes and the products thereof.

Additionally, there was a desire among states not to interfere with agricultural or horticultural cross-breeding processes. This meant that the limitation was also extended to essentially biological processes for the production of plants or animals. This situation reflected states' practice. In this regard also, a claw-back clause was inserted, excluding microbiological products and processes, because micro-organisms had long been used in patented processes.[46] The extension of the exclusion provision to essentially biological processes and the insertion of the claw-back clause highlight the fact that the Strasbourg Convention was drafted to reflect states' practice.

Given that the exclusion provision was *permissive* in that any country with a sufficiently liberal attitude to moral or public order implications was free to omit it, the problem the Convention presented revolved around the wording to be

---

42   Plant Varieties and Seeds Act 1964 UK.

43   The Copyright Act 1956 UK s.10 prevents double protection for designs via copyright and design registration.

44   See the 1961 Act of UPOV Art. 2(1), although this ban has since been removed in the 1991 Act.

45   Strasbourg Convention 1963 Art. 2(b).

46   Including those concerning organic chemicals; see *Commercial Solvents Corporation v Synthetic Products Co. Ltd* [1926] 43 RPC 185.

adopted, particularly in regard to the first exception. That a 'permissive exclusion' was agreed among Contracting States reinforces the argument that the Convention was drafted to accommodate states' practice and suggests that in the context of patentability moral concerns are marginal. This is supported by the observation that there was no uniform practice in national laws refusing patents on the basis of morality for the reason that at the time of drafting there was no consensus among states as to the meaning of morality.

Implementation of the Strasbourg Convention

It was recognized at the time of drafting the Strasbourg Convention that a provision taking the form outlined in Article 2(a)[47] would entail a change in UK patent law. The existing UK statutory provision relating to morality was contained in section 10(1)(b) of the Patents Act 1949, which prohibited inventions the *use* of which would be contrary to law or morality. For the Convention this wording was considered too narrow, since only immoral use would prevent patentability, and not immoral publication.[48] But it was also considered too broad in that patentability could be prohibited where immoral usage would be contrary to law in general. The UK provision was too broad for two reasons. First, something which is contrary to law one year might become legal the next, and, secondly, the fact that use of a product might be proscribed domestically did not necessarily exclude its manufacture for export;[49] hence a proviso in Article 2(a) of Strasbourg. In the final draft the committee decided to adopt the French formula 'Contraire à l'ordre public ou aux bonnes mœurs'.[50] The problem this formula presented to UK patent law was that the term '*ordre public*' was not precisely translatable into English. To resolve such problems another committee was set up in the UK[51] to conduct a review of the British patent system. The terms of reference of the committee included examination of, and reporting on, the British patent system and patent law, in the light of the UK Government's intention to ratify the Strasbourg Convention. This resulted in the Banks Report 1970[52] in which *ordre public* was equated with 'public policy'. The Committee

---

47   *Supra.*

48   Offensive publications could be prevented only by invoking the royal prerogative as outlined in the Patents Act 1949 s 102(1).

49   This principle featured in Article 4 *quater* of the Lisbon Revision 1958 of the Paris Convention. It tells that 'The grant of a patent shall not be refused and a patent shall not be invalidated on the ground that the sale of the patented product or of a product obtained by means of a patented process is subject to restrictions or limitations resulting from domestic law'.

50   Meaning 'Contrary to public order and public morals'.

51   Mr Douglas Jay, then President of the Board of Trade, announced in the House of Commons the intention to set up such a committee.

52   See *The British Patent System, Report of the Committee to Examine the Patent System and Patent Law* (Cmnd 4407 HMSO 1970).

held that *ordre public* means 'public policy', but that it also included 'public order' in so far as this related to rioting, the administration of justice, public services, national economic policy and the proper conduct of affairs in the general interest of the state and society.[53] In order to enable the UK to conform to Article 2(a) of the Convention, it was recommended that: 'The reference to inventions contrary to law in section 10(1)(b)[54] be replaced by a reference to inventions contrary to "*ordre public*" or public policy.'

As already noted, the changes recommended by the Banks Committee were intended to enable UK compliance with the Strasbourg Convention and, by extension, the EPC (the basis of the Patents Act 1977). Not surprisingly, then, the language and concepts of the Patents Act 1977 as understood against the background of British patent decisions are problematic for patent authorities in the UK (Chapter 5).

The European Patent Convention 1973

With the advent of the European Economic Community 1957 (henceforth EEC),[55] it became necessary to harmonize national law among the Member States. Removal of trade barriers was essential for the smooth operation of the Common Market.[56] It was obvious that patent law, with its emphasis on territoriality, could impede the functioning of the Common Market. However, it was also recognized that harmonization of national patent law would not take away from the fact that patents granted would be national. What was required, therefore, was a European patent governed by Community law and dealt with by Community institutions.

A Committee to draft a Community patent was set up.[57] The aim of the Committee was to study the possibility of harmonizing or unifying the laws on industrial property between the six Member States of the EEC.[58] Three Working Groups were nominated initially: patents, trade marks, and designs and utility models. The Working Group for patents[59] was nominated in the beginning of 1960 and, with the approval of the Coordinating Committee, published its draft on 9

---

53   See the Report, *supra* at para. 242.

54   Patents Act 1949.

55   Established by the Treaty of Rome 1957.

56   Community Committees on trade marks and designs were set up, chaired respectively by de Haan (The Netherlands) and Roscioni (Italy), with a view to removal of non-tariff barriers to trade.

57   Chaired by Kurt Haertel of Germany.

58   France, Germany, Italy, Belgium, the Netherlands and Luxembourg.

59   The Working Group consisted of Haertel, President, Germany; along with from one to three experts from each of the six EEC States as well as representatives from both the European Atomic Energy Community (EURATOM) and the European Coal and Steel Community (ECSC).

October 1962, the so-called 'Brussels Text'. In the draft the authors attempted to create:[60]

- a supranational patent system in conformity with the Paris Union;
- a federal patent system coexisting with national systems which had been in force for more than 50 years and often much more than that (as, for example, in the UK);
- a patent system to support the economic structure of the EEC;
- a supranational patent office in a region where at least ten languages prevailed, all with historical backgrounds;
- a compromise between the systems in which patents were granted without search and those which provided an examination as to patentability;
- a patent system holding a fair and equitable balance between the interests of the inventor and those of the public.

Much of the effectiveness of the Committee was due to extensive cross-fertilization taking place between the concurrent Brussels and Strasbourg committees. The aims of the European Patent Convention[61] were largely twofold. First, enforcement of the Treaty of Rome and, second, creation of a supranational legislation for a European patent valid in a greater number of states than members of the EEC.

The legal basis of the first aim was Article 100 Treaty of Rome, which directed the Council to adopt laws effecting the establishment or functioning of the Common Market, and Article 9, which mandated a customs union as the basis of the Community.

More difficult was the second aim, namely, the creation of an international legislation. Here the problem was how to save as much from the existing patent systems as possible in order to render the new system as attractive to as many countries as possible. However, this was not a major problem because the Coordinating Committee directed that the new supranational legislation should coexist with national patent laws. Therefore, the Working Group could create new legislation on its merits without having to grapple with prejudices in existing patent laws. G. Oudemans, a contemporary commentator, wrote:[62]

> The foregoing does not mean that the new legislation was drafted without taking benefit of the experience gathered from the existing patent laws. On the contrary, every separate item was studied against the background of the experts from the six countries since such co-operation could only be beneficial to the ultimate results.

---

60   See generally G Oudemans, *The Draft European Patent Convention: A Commentary with English and French Texts* (1963) Stevens & Sons, London.

61   As it was called, although it was in fact a Community patent system.

62   See generally G Oudemans, *The Draft European Patent Convention: A Commentary with English and French Texts*, *supra*.

This again suggests that the Brussels Text accommodated states' practice.

Impact of Strasbourg

The Strasbourg negotiations were conducted between Council of Europe countries, which included the original six EEC Member States and European Free Trade Area (henceforth EFTA) countries.[63] The obligatory substantive provisions of the Strasbourg Convention were incorporated into the Brussels Text. However, optional exceptions not affording patent protection, allowed for at Strasbourg,[64] could not be incorporated as an option into the Brussels Text. Harmonization within the EEC obliged all Member States to operate the same system of law. In this regard, it was finally decided that the Brussels Text should take up the option of not affording patent protection in respect of certain classes of invention.[65] However, the EEC text was not proceeded with in that form, mainly for political reasons. There was disagreement within the EEC as to whether the text should be confined to Member States only, or extended to non-EEC states also. The idea of a two-convention structure began to emerge, whereby a European patent on the Strasbourg model of substantive law would be open to all countries, but where regulation of the effects of such patents would be restricted to EEC countries.

Documents relating to the Proceedings of the first meeting of the Patents Working Party held at Brussels from 17 to 28 April 1961 outline the discussions underlying Article 12 of the Preliminary Draft regarding exceptions to patentability, in particular inventions whose publication or exploitation would be contrary to *ordre public* or morality.[66] After some preliminary discussion, the Working Party agreed to leave aside for the time being the problem of *ordre public*, and to consider the following problems in the order given:

- Is it necessary to provide for such exceptions to patentability in the European Convention?
- What is the definition of morality? Is there a 'European' definition of morality? Should national definitions be applied or is it necessary to consider what is common to them all?

In an attempt to resolve these issues, the Working Party recognized that there was no European definition of morality and unanimously agreed that interpretation

---

63   Because the European Patent Convention is largely based on the Strasbourg Convention, it is open to non-Community states.

64   Strasbourg Convention 1963 Art. 2(a) and Art. 2(b).

65   Strasbourg Convention Art. 2.

66   See Patents Working Party document: section 5, IV/2767/ 61-E, available from the European Patent Office.

of the concept of morality should be a matter for European institutions.[67] It was, therefore, 'enough to mention the concept of morality in Article 12, without giving further details'.[68] The Working Party approved the text on morality outlined in Article 12 and went on to discuss the exception for *ordre public*. Such was the limited extent of consideration given to the morality provision. This reinforces the view that in terms of the Brussels Text, the basis of the EPC (and the Patents Act 1977 UK), in the context of patentability, morality was marginal.

In 1968 the French Government proposed to its EEC partners the resumption of efforts towards a European patent system. The incentive behind the French initiative was the need for greater protection which 'prior examination' systems conferred.[69] The establishment of a European Patent Office dealing with pre-grant procedure was seen as necessary to achieve this. The French proposal was agreed with other EEC partners, and in March 1969 the EEC Council of Ministers approved a memorandum along similar lines.[70] An Intergovernmental Conference was established for the purpose of setting up a European system for the grant of patents.[71] In accordance with the two-convention approach, progress was made in parallel for both the 'Convention on the Grant of European Patents' and the 'Convention for the European Patent for the Common Market'.[72]

In 1970 the first drafts of the two conventions were published.[73] The Intergovernmental Conference on the setting up of a European system on the grant of patents, at its second meeting in Luxembourg from 13 to 16 January 1970, adopted a first preliminary draft convention which it published.[74] The draft was submitted by the Working Party to the Intergovernmental Conference and was accompanied by a number of reports. There was a general report submitted by

---

67   See Minutes of the Patents Working Party meeting held on 18 Apr 1961, Document IV/2767/61-E, *supra*.

68   Ibid.

69   An aspect inherent in the soon to be completed Patent Cooperation Treaty.

70   See Document BR/2/69 of the Intergovernmental Conference for the Setting Up of a European System for the Grant of Patents, the English version of which is published; see 1970 vol. 1 *International Review of Industrial Property and Copyright Law* 26.

71   Seventeen states were represented at the first meeting held on 21 May 1969: the original six EEC Member States as well as Austria, Denmark, Greece, Ireland, Norway, Portugal, Spain, Sweden, Switzerland, Turkey and the UK. Later Monaco, Yugoslavia, Liechtenstein and Finland also attended.

72   The so-called Community Patent Convention.

73   The European Patent Convention was published in the three languages of the Intergovernmental Conference – English, French and German – while the other Convention was published in the official languages of the EEC at the time, namely, Dutch, French, German and Italian.

74   See Official Publication no. 1350.10.58 on the Intergovernmental Conference for the Setting Up of a European System on the Grant of Patents.

the President and several reports from the various delegations. In the report of the first preliminary draft by the British delegation it was stated, without more, that 'Article 10 (exceptions to patentability) simply follows Article 2 of the Strasbourg Convention'.[75] Since the Strasbourg Convention was concerned with harmonization of obligatory patent law criteria only, there was no attempt made in the draft setting up a European system for the grant of patents to take seriously moral controls as criteria of validity. This supports the view that under the EPC moral considerations are marginal.

In the summer of 1971 second drafts of both conventions were published. In June 1972 the Intergovernmental Conference agreed on a complete draft of the European Patent Convention to be submitted to national governments. In March 1973 the Community Patent Convention was also submitted to national governments. A Diplomatic Conference, convened by the German Government at Munich, agreed the EPC on 5 October 1973. In 1975 the Community Patent Convention was signed in Luxembourg.[76]

As already noted, the Intergovernmental draft European Patent Convention[77] is premised on the Brussels Text of 1962 that was, so far as substantive law provisions were concerned, Strasbourg-based. Additionally, the Strasbourg Article 2 option was taken up in the EPC in Article 53. Armitage and Davies write:[78]

> So far as can be recalled, and so far as official records in this country show, Article 53 was adopted without further discussion as to its purport. It was, in fact, seen in the same light as in 1962 as an unremarkable but necessary marginal safeguard as was conventional in national patent laws as regards Article 53(a),[79] and as a necessary genuflection to the UPOV Convention as regards Article 53(b).[80]

As noted heretofore, documents relevant to the creation of Article 53 EPC support Armitage and Davies and indicate that patentability requirements were never intended to include morality other than as a token operating only on the margins of the system. However, the text of Article 53 EPC, dealing with patentable exceptions, shows a contrary intention and compels exclusion from patentability on moral grounds. Article 53 reads partly as follows:

75　A view noted in the Minutes of the Meeting of Working Party 1, 8–11 July 1969, Doc Br/7/69 at para. 23.

76　Not yet in force.

77　Between 1969 and 1972.

78　See generally Edward Armitage and Ivor Davies, *Patents and Morality in Perspective, supra.*

79　Concerning inventions whose publication or exploitation would be contrary to *ordre public* or morality.

80　Concerning variety-type inventions.

European patents shall not be granted in respect of:

(a) inventions the publication or exploitation of which would be contrary to *'ordre public'* or morality, provided that the exploitation shall not be deemed to be so contrary merely because it is prohibited by law or regulation in some or all of the Contracting States.

Therefore, under Article 53(a) exclusion from patent protection is mandatory for inventions whose publication or exploitation *would be* contrary to *ordre public* or morality.

Additionally, Article 53(b) deals with subject-matter exclusion from patentability. It reads as follows:

European patents shall not be granted in respect of:

(b) plant or animal varieties or essentially biological processes for the production of plants or animals; this provision does not apply to microbiological processes or the products thereof.

Arguably, in two respects Article 53 EPC is a veiled attempt to regulate innovation, including biotechnology: first, by withdrawing the incentive to engage in such activity by denying patents for inventions on moral grounds; and, second, in precluding certain subject-matter from patentability. In this regard, the underlying policy and jurisprudence of Article 53 EPC are examined in Chapter 4.

Morality and UK National Patent Practice

This section briefly outlines, and examines, the approach to morality adopted in the UK where refusal of patents on grounds corresponding to Article 53(a) EPC[81] has its roots in the historic Statute of Monopolies 1624. Before this, granting of patents for inventions occurred through the exercise of the royal prerogative. However, many of the patents so granted were against the common interest. In an effort to curb abuse by the Crown, the Statute of Monopolies was enacted, prohibiting all monopolies except those allowed by virtue of section 6. Under the Act, monopolies were permitted so long as '[t]hey be not contrary to law or mischievous to the State'.[82] The power of the Crown to refuse patents in exercise of the royal prerogative was maintained in successive patent Acts until its final

---

81   EPC Art. 53(a) prohibits patenting of inventions whose publication or exploitation would be contrary to *ordre public* or morality (amended by EPC 2000). See also the Patents Regulations 2000, discussed hereafter.

82   The Statute of Monopolies 1624 s.6.

appearance in the Patents Act 1949.[83] The power of refusal was reinforced in the Patents and Designs Act 1883[84] by giving the Patent Office, in the person of the Comptroller-General, express authority to refuse to grant a patent where '[u]se would, in his opinion, be contrary to law or morality'. The wording of the corresponding section 75 of the Patents and Designs Act 1907 and 1919 was virtually unchanged from that of the 1883 Act.

While there is no reported case of a patent being refused on moral grounds during this period,[85] in *In the Matter for an Application for a Patent by A. and H.*[86] the issue of morality did arise. The application was in respect of 'an improved contraceptive device'. The Assistant-Comptroller[87] refused to accept the application. The applicants appealed to the Law Officer, who is the legal advisor and representative of the Crown.[88] During the hearing the Assistant-Comptroller argued that the Law Officer, who, as representative of the Crown, was charged with the exercise of the prerogative,[89] had ample powers to refuse the grant of a patent for articles not suitable for sale as being protected by Royal Letters Patent. The Law Officer,[90] dismissing the appeal, said:

> The question arises whether ... [t]he Crown in the exercise of its prerogative could possibly be expected to exercise its discretion to grant a patent for an article designed as an apparatus for the prevention of contraception..[I] decline to be any party to the grant of a patent for this class of article ... [E]ven if, as to which I express no opinion, its use as a contraceptive is consistent with morality, I am not prepared to exercise on behalf of the Crown the Crown's discretion in favour of the grant of a patent in respect of it ... [I] express no opinion as to whether the use of these articles is consistent with morality, because I am not aware that the law has laid down what the *exact standards* of morality are. *I am a Court of Law, and not a Court of Morality.*

Two points are worthy of note. First, the Law Officer declared that he was a 'court of law' and not a 'court of morality'. The implication is that morality is removed from the application of patent law, or, at most, is marginal thereto. Second, at the time of the case exact standards of morality were not laid down in the UK. This suggests that uncertainty existed in respect of what constituted morality, so that

---

83 Patents Act 1949 s.102(2).
84 Patents Act 1949 s.86.
85 From Elizabethan times to the Patents and Designs Act 1919.
86 1927 44 RPC 298.
87 Mr Haycraft, acting for the Comptroller.
88 And, in modern practice, the government.
89 Reserved in the Patents and Designs Act 1907 and 1919 s.97. Section 97 stated: 'nothing in this Act shall take away, abridge, or prejudicially affect the prerogative of the Crown in relation to granting any letters patent or to the withholding of a grant thereof'.
90 See the judgment of Sir Thomas Inskip at 298.

uniformity with other national laws was not possible. This reinforces the view it is unlikely for the drafters at Strasbourg to have intended morality as a substantial criterion of patent law ranking with other criteria of validity.

In 1935 the royal prerogative was again used to order the Comptroller to withhold the grant of a patent for a contraceptive. In *In the Matter of an application for a Patent by Rufus Riddlesbarger*,[91] section 97 of the Patents and Designs Acts 1907 to 1932 was in issue. The patent application was entitled 'Improvements in Pessaries'. The device was described in the specification solely with reference to its use for the application of medicaments. However, the Examiner was of the opinion that it was obviously adapted for use as a contraceptive. Consequently, he reported that the grant ought to be refused because the invention belonged to a class for which patents had not hitherto been granted in the UK. An appeal was heard before the Assistant-Comptroller-General. He refused to grant the patent on the grounds that the device was one of a class in respect of which the prerogative of the Crown should be exercised to withhold the grant of a patent under the terms of section 97 of the Act. It is significant that although the Comptroller had power to refuse patents for inventions whose *use* was contrary to morality, both applications were refused in the exercise of the royal prerogative.[92] This suggests that, while the power of the Comptroller to refuse grant was limited to 'immoral use', refusal was (at the time) also permitted under the guise of the royal prerogative where *publication* of the invention was considered undesirable on moral grounds.

The Patents Act 1949 left the provisions dealing with unlawful or immoral use unchanged,[93] and there are no reported applications which were abandoned following an examiner's objection on either moral or public order grounds. Armitage and Davies state: 'Our own recollections of practice in the Patent Office under the Act are that moral considerations focused almost exclusively on sexual morality'.[94] It is thus apparent that the practice in the Patent Office under the predecessors of the Patents Act 1977 focused almost entirely on personal morality.

Impact of EPC on Patent Law in the UK

It was recognized at the time of drafting both the Strasbourg and European patent conventions that a provision taking the form outlined in Article 2(a) and Article 53(a) respectively would necessitate a change in UK patent law. Consequently, the Patents Act 1977 was enacted to give effect to the provisions of the conventions. The Act adopted a formula corresponding to Article 53(a) EPC, compatible with

---

91   1935, 53 RPC 57.

92   By invoking the royal prerogative offensive publications could be prevented; see the Patents Act 1949 s.102.

93   The Patents Act 1949 s.10.

94   See generally Edward Armitage and Ivor Davies, *Patents and Morality in Perspective, supra.*

Article 2(a) Strasbourg, and in accordance with the recommendations of the Banks Committee Report.[95]

The morality provision of the Patents Act 1977 is set out in section 1(3)(a) and (4),[96] which read as follows:

1(3): A patent shall not be granted–

(a) for an invention the publication or exploitation for which would be *generally expected* to encourage offensive, immoral or anti-social behaviour;

1(4): For the purposes of subsection (3) above, behaviour shall not be regarded as offensive, immoral or anti-social only if it is prohibited by any law in force in the United Kingdom or any parts of it.[97]

Section 1(3)(a) has been amended by the Patents Regulations 2000 to read: 'A patent shall not be granted for an invention the commercial exploitation of which would be contrary to public policy or morality'. It is likely that the amendment provision is narrower than the original because the morality provision is triggered by exploitation only. The effect of sections 1(3)(a) and 1(4) of the 1977 Act is much the same in practice as under earlier Acts, although, in some ways, the 1977 Act is broader, in that the morality provision is triggered by publication. By contrast, the 1949 Act was triggered only by immoral use, or use contrary to law. However, as already noted, it was always open under the 1949 Act to control publication in the exercise of the royal prerogative.

The intention behind the morality provision under the current Act, as amended, is summed up by the UK Patent Office's Manual of Patent Practice which reads as follows:[98] 'The effect of s.1(3) remains the same, which is to prevent the grant of patent rights for inventions which the general public would regard as abhorrent or from which the public need protection.' The Manual goes on to tell us that clearly what is to be regarded as contrary to public policy or morality will vary according to changes in social attitudes. While the Examination Guidelines for Patent Applications relating to Biotechnological Inventions in the UK Intellectual Property Office[99] tell us that Section 1(3) of the Act also prevents the patenting of an invention that would be generally expected to encourage offensive, immoral or antisocial behaviour. But this is not what the Act says. Since the effect of

---

95   See the Banks Committee Report *The British Patent System, Report of the Committee to Examine the Patent System and Patent Law, supra* at paras 242 and 243.

96   Amended by the Patents Regulations 2000.

97   This was to accord with the recommendation of the Banks Report, at para. 242, that reference to inventions contrary to law in the Patents Act 1949 s.10(1)(b) be replaced by reference to inventions contrary to *ordre public* or public policy.

98   See the 2008 revision of the Manual at s.1.33.

99   Sept 2007, para. 94.

the amendment remains the same, the conclusion is that the words 'contrary to public policy or morality' are to be equated with the words 'offensive, immoral or antisocial behaviour', meaning that inventions which the general public would regard as abhorrent or from which the public needs protection are not patentable. In terms of morality, whether or not the amendments will result in clarification of standards for interpreting patent rights, and increasing predictability in the application of patent laws in UK national patent practice, remains unclear.

What is clear is that the history of UK patent law involving moral issues is one of scant activity. Cases involving morality largely concerned personal morality only. The morality provision of the Patents Act 1977, invoked to refuse grant of a patent only if the invention concerned is regarded by a large majority of the public as abhorrent or from which the public needs protection, suggests that a 'light' moral regime was what the UK intended. And in this regard, there is no evidence to suggest that the Patent Office ever was, or is now, concerned with taking the lead in resolving difficult moral issues in wider directions. This contrasts sharply with the position adopted by the European Patent Office (Chapter 4).

Conclusions

In drafting the Strasbourg Convention, most of the intellectual effort went into defining, and reshaping, concepts such as novelty, inventive step and industrial character, which were seen as substantial elements of patent law, and, therefore, obligatory. However, because the term 'morality' is not defined and (exact) standards were not laid down in the Convention, it would seem exclusion from substantive criteria for patentability was intentional.

The legislative histories of the Strasbourg Convention 1963 and the European Patent Convention 1973 reveal an intention on the part of the drafters to continue national practice in respect of powers to refuse patents on the basis of morality. National patent practice in the UK suggests that moral considerations were never intended to be a criterion of substantive law. On the contrary, what little case law exists suggests that only those inventions concerned with sexual morality were precluded from patentability. However, because moral norms change with time, sexual morality is no longer the contentious issue it once was for states. In contrast with the European Patent Office, there is no tendency on the part of the Patent Office in the UK to take the lead in resolving other difficult moral issues in wider directions. Indeed, it is the author's contention that all patent-granting authorities should adopt a flexible attitude towards changing standards of morality when reviewing patent applications, since, once a patent application is refused, it cannot be revived.

Because the morality provision is presented as an exception to patentability, this suggests that it should be interpreted in a restrictive manner. And meaningful interpretation of law requires adaptation to actual circumstances. In the past, this meant the call from industry for harmonization and centralization was answered by states agreeing among themselves the substantive criteria for patentability. In the future, in the interests of certainty and clarity, it may mean states having to decide whether morality is a criterion for patentability or merely marginal.

Chapter 3

Development of Patent Law in the United States

Introduction

The purpose of the US patent system is to promote work in the sciences and useful arts by providing a reward to inventors as an incentive to disclose information for the benefit of the public. The reward is given in the form of a grant from the federal government of the right to exclude others from making, using or selling the invention. Prima facie, therefore, the US patent system is akin to the European system. However, as already noted, there are differences resulting in innovators in the US receiving greater protection than their European counterparts. US public policy regards patents as a means for creating new industries and jobs. The US patent system is based on a quid pro quo: strong protection is granted for a limited time in exchange for complete disclosure. By granting an exclusive right to the inventor to manufacture and sell the invention for a period of time, the inventor is offered a financial incentive to invest in the discovery of new innovation. The incentive is particularly important in the pharmaceutical and biotechnology fields, where meaningful discoveries often require extremely expensive and time-consuming research.[1]

Origin of Protection

The grant of power to Congress to enact laws relating to patents is found in Article 1, section 8 of the Constitution, which reads:

> The Congress shall have power–

> To promote the progress of science and useful arts, by securing for limited times to authors and inventors the exclusive right to their respective writings and discoveries.

---

1   Jacqueline D Wright, 'Implications of Recent Patent Law Changes on Biotechnology Research and the Biotechnology Industry', 1997 vol. 1 *Virginia Journal of Law and Technology* 1–13 at 1.

The provision provides separately for copyright legislation to secure protection for 'authors' over their writings with the purpose of promoting 'science', and for patent legislation to secure protection for 'inventors' over their 'discoveries' with the purpose of promoting 'useful arts'.

The clause is both a grant of power and a limitation.[2] The qualification is limited to promoting advances in the 'useful arts'. Congress in the exercise of the patent power may not overreach the restraints imposed by the stated aim of the Constitution. Neither can it enlarge the scope of monopoly without regard to the innovation gained thereby. Innovation, advancement and things adding to the sum of useful knowledge are inherent requisites in a patent system, which, by constitutional command, must promote the 'useful arts'. Within the limits of the constitutional grant, Congress may implement the stated purpose of the framers by selecting a policy which, in its judgment, best effectuates the constitutional aim. Congress responded quickly to the framers' intent and enacted the Patent Act 1790, giving rise to the utility patent.[3]

In this regard, an agency[4] was created in 1789 within the Department of State and headed by the Secretary of State, the Secretary of the Department of War and the Attorney General, any two of who could issue a patent for a period not exceeding 14 years. A petitioner who 'hath invented or discovered any useful art, manufacture, or device or any improvement therein not before known or used' and who satisfied the agency or board that the invention or discovery was sufficiently useful[5] and important was issued with a patent.

Thomas Jefferson in 1789, who as Secretary of State was a member of the Board, might justifiably be called the first administrator of the patent system.[6] However, Jefferson had an instinctive aversion to monopolies[7] and did not believe in granting patents for small details or obvious improvements. Jefferson had difficulty in drawing a line between innovation over which an exclusive patent should be granted in the public interest and that which should not. In this regard, the Board formulated several rules but it was recognized with what slow progress such a system of rules could mature. The whole matter was turned over to the judiciary, to be matured into a system, under which all patentees might know

---

2   See generally, Donald S Chisum, *Patents: A Treatise on the Law of Patentability, Validity and Infringement* (1978) Matthew Bender & Co., NY. Updated periodically, most recently Sept 2000.

3   Ibid.

4   The Agency or Board was known as the 'Commissioners for the Production of the Useful Arts', created by the Constitutional Convention of 1789.

5   Discussed more fully hereafter.

6   Jefferson was chiefly responsible for implementing the first Patent Act in 1790. See generally, Robert Patrick Merges, *Patent Law and Policy-cases and Materials* 2nd edn (1997) Michie Law Publishers, Va.

7   Ibid.

when actions were lawful. Congress agreed with the Board that the courts should develop additional conditions for patentability.[8]

Although the Patent Act was amended, revised or codified some 50 times between 1790 and 1950, Congress steered clear of a statutory set of requirements other than the bare novelty and utility tests reformulated in Jefferson's draft of the Patent Act 1793. The initial Patent Act 1790 covered 'any useful art, manufacture, engine, machine or device, or any improvement thereon', and the 1793 Act was altered to read 'any art, machine, manufacture or composition of matter, or any new and improvement thereon'. Thereafter the statutory classes remained the same until 1952 and all the while the courts consistently interpreted 'art' as 'process'. To clarify the situation, the Patent Act 1952 changed the language from 'art' to 'process', and defined 'process' as meaning 'process, art or method'.

While the first patent law was enacted in 1790, the law now in effect is a general revision enacted 19 July 1952, which came into effect on 1 January 1953. It is codified in Title 35, United States Code (hereafter USC).

The difficulty of formulating conditions for patentability was highlighted by the generality of the constitutional grant and statutes implementing it. Utility and novelty as defined in section 101 and section 102 of the current statute trace closely the 1874 codification and express the 'usefulness and new' tests which have always existed in the statutory scheme. However, implementation of the public policy that 'things which are worth to the public the embarrassment of an exclusive patent'[9] was more problematic.

In *Hotchkiss v Greenwood*[10] (henceforth *Hotchkiss*) in 1850, the Supreme Court formulated a general condition for patentability. The court said: 'Unless more ingenuity and skill were required than were possessed by an ordinary mechanic acquainted with the business, there was an absence of that degree of skill and ingenuity which constitute essential elements of every invention.' The *Hotchkiss* test laid the cornerstone of the judicial evolution suggested by Jefferson and left to the courts by Congress and 'invention' as a term of legal art came to mean patentable invention. In this manner, the Supreme Court laid down the basis for section 103, Title 35, USC, namely, non-obvious subject-matter.

The four classes of statutory subject-matter (i.e., any art, machine, manufacture or composition of matter) have been called the 'great and distinct classes of invention'.[11] By contrast, in Europe under the European Patent Convention (hereafter EPC), Article 52 outlines classes of invention not patentable. Article 52(1) EPC provides generally that patents 'shall be granted for any inventions which are susceptible of industrial application, which are new and which involve an inventive step'. What amounts to an inventive step has, in the past, been

---

8 See *Graham v John Deere Co of Kansas City*, decided 21 Feb 1966, 148 USPQ 464.

9 As Thomas Jefferson put it. See generally Robert Patrick Merges, *Patent Law and Policy-cases and Materials* (1997), *supra*.

10 (1850) 11 How 248.

11 *Ex Parte Blythe* 1885 Commission Dec 1862, 86 (Comm'n Pat 1885).

difficult for courts, especially in the UK.[12] Additionally, the possibility exists that one day there may be a case 'in which it is necessary to decide whether something which satisfies the conditions can be called an invention, but that question can wait until it arises'.[13] 'That question' is likely to arise in biotechnology where even the smallest advance in the art can amount to an inventive step. In contrast to the US, the issue of non-obviousness is not resolved within the context of the EPC. Article 52(2) outlines four classes of non-patentable invention, while Article 52(3) states that patentability is precluded 'only in so far as the patent relates to such subject-matter or activities as such'. Interpreting Article 52(3) has also been problematic for courts in the UK. Article 52(4)[14] tells that diagnostic and therapeutic methods of treatment of the human or animal body are not regarded as inventions susceptible of industrial application.[15] The question arises: on what basis are the four classes of subject-matter excluded from patentability by virtue of Article 52(2) EPC?

The general purpose of the statutory classes of subject-matter in the US is to limit patent protection to the field of applied technology, that is, the 'useful arts'. In this regard, the four classes are quite flexible and have been interpreted to cover new technologies that evolved over the past two centuries (Chapter 6). It seems that enforcement mechanisms in the US, facilitated by positive and clearly drafted legislation, permit judicial legislation in a manner the EPC does not.

As a result of the importance of technology in a competitive economy, the US has striven to ensure that its patent system continues to maximize the incentives for innovation and development. The past success of its patent system in providing this stimulus is a direct result of its ability to evolve and adapt to changing times and new technologies as they arise. By contrast, under the EPC patent law policy is concerned largely with the creation and enforcement of the internal single market.

The US has sought to harmonize the patent laws of different nations and to establish minimum standards for acquisition and enforcement.[16] Perhaps the most significant legislative reform has been the creation of the Court of Appeals for the Federal Circuit in 1982 (henceforth CAFC).[17] The CAFC was created to promote

---

12   See Ch. 5 hereafter.

13   See *Biogen Inc. v Medeva PLC* [1997] RPC 1, per the judgment of Lord Hoffmann.

14   Since EPC 2000 (in force 2007) the contents of Art. 52(4) have been transferred to Art. 53(c) EPC.

15   Very little literature exists relating to the exclusion of methods of medical treatment from patent protection, and especially so in regards to animals. See generally Eddy D Ventose, 'Farming Out an Exemption for Animals to the Method of Medical Treatment Exclusion under the European Patent Convention', 2008 vol. 30 *European Intellectual Property Review*, esp. at 509.

16   See generally *The Advisory Commission on Patent Law Reform* report to the Secretary of Commerce (Honourable Barbara Hackman Franklin) (Aug 1992) 1–27, published by *The World*, Public Access Unit, Brookline, Mass.

17   The CAFC was created on 1 Oct 1982 by the merger of two existing Art. 111 constitutional courts. These were the Court of Claims and the Court of Customs and

greater uniformity in the application of patent law and to reduce the possibility of forum shopping by parties seeking favourable courts.[18] The CAFC now has exclusive jurisdiction over all appeals in patent cases. From its creation, the CAFC has been decidedly 'pro-patent'.[19] However, whether this is or is not a good thing is questionable;[20] the result is that patent protection in the US has increased and patent litigation has become increasingly worthwhile. What the CAFC has done is to provide much-needed clarification of standards for interpreting patent rights and increasing predictability in the application of the patent laws. Clearly, such a development is welcome. By contrast, because the European Patent Office is not an EU institution, legislative reform of this nature is not possible.

Moral Issues and US Patent Law

Courts in the US were often willing to withhold patents for inventions they considered immoral. Historically, these fell into two classes: inventions used to defraud buyers and machines used for gambling. Although US patent law has no morality clause per se, because moral norms were often enforced in the courts by means of the utility requirement, the link between 'utility' and 'morality' is important and outlined briefly hereafter. The US concept of utility is both broader and narrower than the notion of industrial application under the EPC.[21] It is a broader concept in that the same word encompasses elements such as morality and illegality. It is narrower in that pure research is not held to equal a practical utility.[22]

Utility has long been a requirement for patentability in American law.[23] Title 35, USC, in section 101, requires that a process, machine, manufacture or composition of matter or improvement thereon be 'useful'. The Patent Act separately requires in section 112 that the specification of a patent application contain a written description of the invention and of the manner and process of making and 'using'

Patent Appeals.

18   See Jacqueline D Wright, 'Implications of Recent Patent Law Changes on Biotechnology Research and the Biotechnology Industry', *supra*.

19   Ibid.

20   The assumption that patent protection incentivizes innovation has never been convincingly demonstrated; an explanation for the increase in patenting is the fact that the establishment of complex patent portfolios is increasingly becoming a business strategy. See generally Michael Blakeney, 'The Role of Competition in Bio-technological Patenting and Innovation', 2006/2007 vol. 9 *Bio Science Law Review* esp. at 95.

21   See Margaret Llewelyn, 'Industrial Applicability / Utility and Genetic Engineering: Current Practices in Europe and the United States', 1994 vol. 11 *European Intellectual Property Review* 473–80 at 474.

22   *Brenner v Manson*, *In re Joly* and *In re Kirk* all established that a compound that is only useful in research does not meet the utility criterion even if that research could lead to a useful product.

23   Patent Act 1793.

it, that is, an enabling disclosure of how to use the invention is required. On 5 January 2001 the US Patent and Trademark Office (henceforth PTO) published utility examination guidelines for its examiners.[24] It should be noted that these guidelines do not have the force and effect of law and make no mention of morality or public policy issues. Under the guidelines, utility encompasses elements of 'specific, substantial, and credible'.[25] What this means is, if the applicant has asserted that the claimed invention is useful for any particular practical purpose, it has a 'specific and substantial' utility; and if the assertion would be considered credible by a person of ordinary skill in the art, a rejection based on lack of utility should not be imposed.[26] In light of the revisions to the US utility guidelines, whether or not moral considerations will continue to be enforced by means of the utility requirement is uncertain.

To be useful, an invention must be capable of some beneficial use in society.[27] The principle was judicially laid down in a pair of cases in 1817, prior to the establishment of the PTO. In *Lowell v Lewis*,[28] Justice Story said: 'All that the law requires is, that the invention should not be frivolous or injurious to the well-being, good policy, or sound morals of society. The word "useful", therefore, is incorporated into the Act in contradistinction to mischievous or immoral.' Justice Story enumerated examples of immoral inventions including a new invention to poison people, or to facilitate private assassination, or to promote debauchery. In *Bedford v Hunt*[29] the same Justice Story said: 'by useful invention, in the statute, is meant a one that may be applied to some beneficial use in society ... [t]he law does not look to the degree of utility; it simply requires that it be capable of use'. The Story view of utility has generally been accepted by the courts and applied in cases involving fraud and gambling devices. In early cases it was a commonly held judicial view that an invention the only use of which is to deceive or commit fraud lacked utility. *Richard v Du Bon*[30] involved a process for artificially producing spots on tobacco leaf. The quality of the leaf was not improved, but was made to look like leaf of a superior quality. The court held the invention unpatentable for lack of utility.

---

24   Federal Register, vol. 66 no. 4, 5 Jan 2001; Examination Guidelines for the Utility Requirement.

25   In *Fisher v Lalfudi* (2005) 04-1464, the CAFC considered nine principles for assessing the utility of biotechnological inventions. See generally Robert Fitt and Edward Noddler, 'Specific, Substantial and Credible? A New Test for Gene Patents', 2008 vol. 9 *Bio Science Law Review* esp. at 183.

26   Examination Guidelines for the Utility Requirement at Part B, point 3.

27   See Richard H Kjeldgaard and David R Marsh, 'Recent Developments in the Protection of Plant-based Technology in the United States', 1997 vol. 1 *European Intellectual Property Review* 16–20 at 16.

28   15F Cas 1018 (no. 8568), Circuit Court, Massachusetts 1817.

29   3F Cas 37 (no. 1217), Circuit Court, Massachusetts 1817.

30   103f 868 (2D Cir 1900).

Unfortunately, however, Justice Story could not anticipate the wonders of modern science. In *Re Bremner*[31] the application did not make any assertion of utility for the claimed compound and the court affirmed the Patent Office Examiner's rejection of the claims for failure to include an assertion of utility in the patent application. Ten years later the CCPA made an about-face on the requirement for a specific assertion of utility in a patent application.[32] Indeed, the court held that a steroid compound could be useful as a building block of value to the researcher. The modern rule of utility came from an appeal to the Supreme Court. In *Brenner v Manson*[33] the Supreme Court considered the sufficiency of the utility requirement. The claim related to a process for producing a steroid said to have tumor-inhibiting effects in mice. Manson argued that a chemical process is 'useful' within the meaning of section 101 if 'it worked' or belonged to a class of compound the subject of serious scientific debate.[34] The court rejected the argument. The court took the view that producing a product belonging to an important 'class' of compounds, namely, steroids, is not sufficient to meet the utility requirement. The court said: 'a *process* to produce a compound may be patented only if the compound has substantial utility or specific benefit ... [i]n currently available form'. The function or utility of the actual steroid produced is required. The decision is premised on the logic that until a process claim has been reduced to production of a product shown to be useful, the bounds of monopoly are not capable of precise delineation. If such a patent were granted, whole areas of scientific development might be monopolized without compensating benefit to the public.[35]

However, merely because an invention is deceptive does not necessarily suggest it lacks utility and the views of the judiciary on such matters have changed with time. In a recent decision, *Juicy Whip Inc. v Orange Bang Inc. and Unique Beverage Dispensers Inc.*[36] (henceforth *Juicy Whip*),[37] the issue of deceptive inventions arose. The district court held the patent invalid for lack of utility on the ground that the patented invention was designed to deceive customers by imitating another product and thereby increasing sales of a particular good. The CAFC reversed. In delivering the opinion of the court, Judge Bryson noted that the requirement of utility is not a directive to the PTO or the courts to serve as arbiters of deceptive trade practices. Other agencies, such as the Federal Trade Commission or the Food and Drug Administration, are assigned the task of protecting consumers from fraud and deception. He said: 'Congress never intended that the patent laws

---

31   182 F. 2d 216 (CCPA 1950).

32   *In re Nelson* 182 F.2d 217 (1960).

33   148 USPQ 689 [1966].

34   Steroids, in this case.

35   A similar view was expressed in *Biogen* [1997] RPC 1 per Lord Hoffmann; see generally Ch. 4.

36   185F 3d 1364, 1999.

37   Decision of 6 Aug 1999.

should displace the police powers of the States, meaning by that term those powers by which the health, good order, peace and general welfare of the community are promoted.' This suggests that moral concerns traditionally within patent law are now addressed elsewhere and there is no need for patent law to be involved in this way. In light of the decision in *Juicy Whip*, inventions used to defraud buyers, whether or not moral, can no longer be considered precluded from patentability on such grounds.

Early American cases took a strict view of patents on gambling devices. If the actual use of the invention was to encourage gambling, despite other potential uses, patentability was denied. In *National Automatic Devise Co. v Lloyd*[38] the court held unpatentable a 'toy automatic race horse'. The court found that the only use of the device to date was 'in saloons, bar-rooms, and other drinking places, where the frequenters of such places make wagers as to which of the toy horses will stop first'. In later American cases a more flexible and sensible approach was adopted. *Koppe v Burnstingle*[39] concerned a patent for a game called 'golf dice'. The outcome of the game was determined by chance and, therefore, the game could be used for gambling or amusement purposes. The court held that the patent could not be held invalid merely because the device of which the patent was descriptive might be put to an illegal purpose. More recently, in *Re Murphy*[40] involving a claim to a 'one-armed bandit' disclosed solely as a gambling device, the court did not find any basis in Title 35, USC, section 101 or related sections justifying a conclusion that inventions, useful only for gambling, are void of patentable utility. In *Juicy Whip*, Judge Bryson went further and said:

> The principle that inventions are invalid if they are principally designed to serve immoral or illegal purposes has not been applied broadly in recent years. For example, years ago courts invalidated patents on gambling devices on the ground that they were immoral, *but that is no longer the law.*

In light of the *dictum* in *Juicy Whip*, inventions used for gambling purposes are no longer precluded from patentability on moral grounds.

As already noted, courts have talked in general terms of morals, health and good order of society. However, moral standards change in a period of a few years.[41] Courts in determining 'utility' based on public mores should apply a test which will not penalize the inventor who may be prescient enough to have anticipated basic needs of a society changed by forces yet unrecognized by the general public.

---

38   40 F 89 (ND Ill. 1889).

39   29F 2d 923 (DRI 1929).

40   200 USPQ 801 [1977].

41   Gambling devices frowned upon early in the twentieth century are now legal in several US states, as are birth-control devices.

Although, in the past, courts have gradually rejected a 'utility' based on public mores for traditional mechanical-type inventions, the cases suggest the link between 'utility' and 'morality' is especially relevant for modern biotechnological inventions involving gene sequences not identified in terms of structure or function.

Patent claims to DNA fragments present special problems for the practical utility requirement under section 112. *Ex parte Tanksley*[42] involved claims relating to the tomato chromosome map and cDNA library. The applicants claimed to have identified the position of cDNA clones on the chromosome map but did not identify any specific use for such clones. The issue was: did the claims comply with the requirement of section 112? The second paragraph of the section reads: 'The specification shall conclude with one or more claims pointing out and distinctly claiming the subject-matter which the applicant regards as his invention.' The Board of Patent Appeals and Interferences held that, when reviewing applications for compliance with section 112, the examiner can require the invention to be described in terms sufficiently definite to enable comparison with available art, meaning that claims must be described either in terms of structure or function. This is justified on the quid pro quo principle, namely, for the purposes of indicating to the public the extent of the subject-matter to which the exclusionary right granted by the patent is intended to apply, comparison with prior art is necessary. The Board held that since none of the cDNA clones claimed had been identified by structure, and only a very few by function, there was no compliance with section 112.

Ex parte Thomas F Deuel[43] involved claims relating to purified growth factor isolated from prostate tissue and to a method of stimulating the growth of fibroblast cells using the growth factor. The Board of Patent Appeals and Interferences considered whether or not the application complied with section 101 utility, and section 112 how-to-use, requirements. The Board noted that, because there was no statement of use in the specification, no description in the specification regarding how to use the claimed growth factor, and no exemplification of a use, the application failed under both section 101 and section 112.

Given the Board of Appeal's decisions in these cases, the question arises in respect of biotechnological inventions, whether or not there is a veiled attempt in the US to enforce moral norms by means of the utility requirement. In response to a recent application, the Newman Application, for a patent on a method for making creatures that are part human and part animal,[44] a PTO press release said: 'It is the position of the PTO that inventions directed to human/non-human chimera could, under certain circumstances, not be patentable because, among other things, they would fail to meet the public policy and morality aspects of the utility requirement.'

---

42   26 USPQ 1348, decided 10 Oct 1991.

43   27 USPQ 1360, decided 28 Jan 1993.

44   See *In Vahalla* applicant (Stuart A Newman, New York Medical College), published in the *Washington Post*, 2 Apr 1998 A 12, article entitled 'Patent Sought on Making of Part-human Creatures', by Rick Weiss.

Basically, the Newman Application was not favourably looked upon because it 'embraced' a human being, or put another way, the hybrid was 'too-human' and thus, did not constitute patentable subject-matter.[45] However, by issuing the press release the PTO showed that it is willing to continue to rely on Justice Story's formulation of utility. The extent to which the press announcement will operate to revive the moral utility doctrine is uncertain given that the Examination Guidelines for the Utility Requirement 2001 make no mention of morality or, indeed, public policy issues. Nevertheless, in this regard Newman's effort to win a patent was only finally rejected by the PTO on 11 February 2005. The grounds for refusal included lack of a written description and lack of utility. Newman asserted ten utilities and the Examiner rejected all of them for not being specific, substantial or credible. One rationale for refusal is that such a patent would be inconsistent with the constitutional right to privacy; another is the possible conflict with the Thirteenth Amendment – prohibition against slavery. Since the Examiner rejected the utilities asserted by Newman, it would have been consistent to use the moral utility doctrine. This did not happen. The conclusion is the moral utility doctrine has fused with the rules regarding patentable subject matter. This would exclude Examiners from making moral judgments regarding what should be patentable subject-matter, and requiring a broad interpretation of what is patentable. It is the author's contention that, despite the PTO decision, the place of moral issues in US patent law is very limited.

Conclusions

The constitutional origin of intellectual property law in the US enables the judiciary there to embark on judicial legislation. Criteria for patentability are laid down by the judiciary suggesting that, in the US, traditional patent law criteria are malleable and, in contrast to European *fora*, readily adapted to meet the needs of new technology. The result is that biotechnological inventions in the US receive greater protection than their European counterparts. In addition, morality as a criterion of substantive validity is not included in the patent laws of the US and technological inventions receive protection once they conform to traditional patentability criteria only, without difficult assessment of moral considerations.

The wording of the Constitution permits Congress to enact legislation in clear and positive terms. Legislative reform in the US has resulted in the creation of the Court of Appeals for the Federal Circuit with exclusive jurisdiction over all appeals in patent cases. This promotes greater uniformity in the application of patent law and is welcome.

The utility criterion and the manner in which morality is used in the courts suggest that moral concerns within patent law have now been addressed in other

---

45   See 'Patent Application Is Disallowed as "Embracing" Human Being', 1999 vol. 58 *Patent, Trademark and Copyright Journal* at 203.

legislation. However, aspects of some recent biotechnology cases could be taken to indicate a limited place in the patent system for moral norms, suggesting that such considerations, albeit residual, continue to apply.

Chapter 4
Application of Article 53
European Patent Convention

Introduction

While developed initially in the area of lower organisms, biotechnological inventions now are increasingly applied to more complex biological entities.[1] To help keep society in step with modern developments in biotechnology, the legal system must respond to newly raised issues in an adequate manner. In particular, risks and undesirable applications of the technology must be avoided or minimized, and great efforts should be made to use the positive potential of the technology fully.

Biotechnology raises many issues previously unknown. Developments in human genetics offer a clear example.[2] Will it be possible to confine the results of viable genetic testing to those who are directly parties to a particular test? If so, will certain genetic profiles then be read as 'good' or 'bad' risks? Will genetic testing be seen as a basis for new classifications, giving rise to new possibilities for discrimination by third parties? In this regard, the problems presented by modern genetics will not be limited to fine points about patentability.

How will biotechnology impact on patent law? Under the revised version of the European Patent Convention (EPC 2000),[3] the response of the legal system to issues associated with biotechnology (and other new technologies) is to withdraw the incentives to engage in such activity by continuing to deny patents for inventions on moral and *ordre public* grounds (outlined in Article 53(a)) and in respect of the technology per se to preclude certain subject-matter from patentability (outlined in Article 53(b)). In addition, under the revised EPC,[4] methods of treatment or diagnosis of the human or animal body are no longer to be treated as unpatentable due to a lack of industrial applicability; now they will lack patentability in their

1 See Rainer Moufang, 'Patentability of Genetic Inventions in Animals', 1989 vol. 20 *International Review of Industrial Property and Copyright Law* 823–46 at 823.

2 See R Brownsword, W R Cornish and M Llewelyn, Editors comment: 'Human Genetics and the Law: Regulating a Revolution', 1998 vol. 61 *Modern Law Review*.

3 The revision was occasioned primarily by the amendment of the Articles agreed at the Diplomatic Conference in Nov 2000. In force 13 Dec 2007.

4 A good summary of the revised EPC 2000 and the potential impact of these changes on European patent practice is given by Stacey J Farmer and Dr Martin Grund, 'An Overview of the New European Patent Convention (EPC 2000) and Its Potential Impact on European Patent Practice', 2006/2007 vol. 2 *Bio Science Law Review* at 53.

own right (outlined previously in Article 52(4)). Moving the exclusion to Article 53 suggests that, in reality, therapeutic and diagnostic methods of treatment are excluded from patentability for public policy reasons.

As already noted,[5] in Europe under the EPC the root of intellectual property law is national and the right to regulate exploitation resides with national authorities. Although a form of the Community patent is 'imminent' (Chapter 9), the concept of 'European patent', really a bundle of national patents, predominates by virtue of the EPC.[6]

By contrast, in the United States the root of intellectual property law is constitutional, the aim being to promote the progress of science and the useful arts.[7] Congress has power to legislate to achieve this aim. Congress also has the right to regulate commerce. In this manner a distinction is drawn between technology per se and its use. This suggests that in the US there is a distinction between patenting the invention and commercialization of results. There are structural variations also. As already noted, in the US the structure of intellectual property law is all-embracing to the extent that the patent system there does not contain a list of excluded subject-matter[8] (discussed more fully in Chapter 6 hereafter). By contrast, the EPC patent system does contain a list of excluded subject-matter.[9]

Arguably, the effect of Article 53 EPC is to *permit* patent law, an instrument conceived with the aim of promoting technology, to reshape itself in such a manner that it no longer reflects, or reflects adequately, the economic concerns surrounding biotechnology. In addition, in the light of the International Union for the Protection of New Varieties of Plants 1991 Act (henceforth UPOV 1991), the logic governing the policy of Article 53(b) EPC requires examination. UPOV 1991[10] removed the 'double bar' on protection, so that now a state can, if it wishes, grant and protect breeders' rights by means of a patent.[11] Arguably, if Article 53(a) were to be abolished it would also be inappropriate to retain Article 53(b), for the following reasons. First, Article 53 EPC is expressed in parts suggesting a link between them. Since, in the text of the Convention, Article 53(b) follows Article 53(a), the expectation is that (b) is of lesser importance than (a); hence abolition of the more important suggests abolition

---

5   See generally Ch. 2 heretofore.

6   Convention on the Grant of European Patents, 5 Oct 1973 Munich, in force Oct 1978.

7   Constitution Art. 1, s.8.

8   Title 35 USC s 101 deals with subject-matter eligibility for patent protection, and there is clearly favour in granting protection of industrial property over life forms. See also Angus J Wells, 'Patenting New Life Forms: An Ecological Perspective', 1994 vol. 3 *European Intellectual Property Review* 111–18 at 112.

9   Excluded subject-matter is to be distinguished from inventions which are not patentable inventions within the meaning of the Convention: see Art. 52(2) EPC.

10   In force 21 Apr 1998.

11   1991 Act Art. 2 states: 'Each Contracting Party shall grant and protect breeders' rights'.

of the other. Second, the exclusion of animal varieties in subsection (b) is also a concern about morality and it is logical that the exclusion outlined in Article 53(a) applies equally to the categories of excluded subject-matter outlined in Article 53(b). In other words, also excluded from patentability on the basis of morality are 'varieties' and essentially biological processes for the production of plants or animals as outlined in Article 53(b). Third, subject-matter limitation is outlined in Article 52(2) EPC,[12] suggesting that subject-matter limitation is not the purpose of Article 53(b).

In national patent law and under the EPC, judges and lawyers are reluctant to address legal issues in moral terms suggesting that Article 53(a) is difficult to enforce. And because Article 53(b) EPC, a complex and negatively drafted legislative provision, is also a concern about morality, this suggests that it too is difficult to enforce. Strict construction of Article 53(b) by the European Patent Office (henceforth EPO) not only ensures that the law is unclear to breeders, but perpetuates a dual system of protection for living material, namely, patent and plant variety protection. What is required is a scheme providing greater encouragement to animal breeders developing breeds through genetic engineering methods by covering the gaps in protection that currently exist.[13] One problem presented for biotechnological inventions is that, at present, animal variety-related inventions receive no protection at all.[14]

This chapter examines EPO jurisprudence in the light of the policies underlying Article 53 EPC and assesses whether or not such policies continue to apply. In respect of Article 53(a), the implications of EPO jurisprudence, according to which morality is a matter for European institutions, is considered in the context of national law. In respect of Article 53(b), case law developments, suggesting that judicial authorities[15] deny patentability to a wide range of biotechnological inventions, are assessed.

Application of Article 53(a)

The practice of the EPO indicates that it is unlikely the morality provision is intended to have any widespread *use* as a bar to patentability. Nevertheless, in this

---

12   This outlines what, in particular, shall not be regarded as invention within the meaning of paragraph 1, and refers *inter alia* to discoveries, aesthetic creations, programs for computers and presentations of information.

13   See Nicholas Peace and Andrew Christie, 'Intellectual Property Protection for the Products of Animal Breeding', 1996 vol. 4 *European Intellectual Property Review* 213–33 at 233.

14   See R Stephen Crespi, 'Prospects for International Cooperation', in William Lesser (ed.), *Animal Patents: The Legal, Economic and Social Issues* (1989) 30–38 at 35, Macmillan, UK.

15   Meaning patent-granting authorities, in particular the EPO.

regard the EPO is empowered. In the interests of certainty and clarity in the law, and because patentees are entitled to know when their actions are safe and lawful, the application of Article 53(a) in a uniform and consistent manner is necessary to protect adequately biotechnological inventions. The cases discussed hereafter demonstrate that the EPO is unclear in its guidance for future cases, suggesting that Article 53(a) is not an appropriate mechanism for determining exceptions to patentability. Based on states' practice, what is required is a definition of *ordre public* and morality concerns that can be controlled by specialized agencies.

As already noted, Article 53(a) EPC deals with inventions unpatentable on the basis of *ordre public* or morality. It reads as follows:

European patents shall not be granted in respect of:

(a) inventions the commercial exploitation of which would be contrary to '*ordre public*' or morality, provided that the exploitation shall not be deemed to be so contrary merely because it is prohibited by law or regulation in some or all of the Contracting States.

The Guidelines for Examination first published by the EPO in 1977[16] outline the policy underlying the interpretation of Article 53(a). It states:[17] '[a] fair test to apply is to consider whether it is probable that the public in general would regard the invention as so abhorrent that the grant of patent rights would be inconceivable'. However, the Guidelines are unclear. They are silent on how the attitude of the 'public in general' is ascertained. Should surveys or opinion polls be taken into account?[18] Should the attitude of parliaments in the Contracting States be considered? Should the attitude of the European Parliament be decisive?[19] The lack of clarity in the EPO Guidelines supports the view that it was never, and is not now, the intention of the drafters of the EPC to permit European institutions to determine patentability using moral criteria on anything more than a cursory basis. That the Guidelines are unclear is testament to the fact that there exists no single European concept of morality. What the drafters of the EPC did envisage was that an *individual state* might refuse patentability on the basis of what it

16   Updated in July 1999, Oct 2001, Nov 2003 and most recently in Dec 2007.
17   The Guidelines for Examination of the European Patent Office (2007), Part C, Ch. IV, s.4.
18   In the *PGS* case [1995] EPOR 357, the TBA rejected suggestions that either surveys or opinion polls were probative of public opinion. See the Reasons for the Decision at point 15.
19   The European Parliament adopted Resolutions declaring that 'transgenic animals engineered to suffer' should be non-patentable on the basis that the production of such animals is contrary to 'public order'. Resolutions B3-0199, 0220 and 024/93, O J E C 1993 C, 72/127.

considered to be an immoral invention.[20] This suggests that the EPO is limited to invoking Article 53(a) in a 'light' manner only. In the context of harmonization of patent law and practice, the fear is that in national patent law and under the EPC the morality provision is open to different interpretations. If so, it is likely that the supranational EPO jurisprudence will be dominant. Consequently, EPO decisions will, ultimately, map out a 'moral frontier' and define what amounts to morally acceptable technological development. This, however, is the job of governments and not of patent-granting authorities. The application by the EPO of Article 53(a), and whether or not it is an appropriate mechanism for determining exceptions to patentability, is examined in this section.

Animal Genetics and Article 53(a)

In the field of animal genetics the case of *Harvard/Onco-Mouse*[21] (henceforth *Onco-Mouse*) is paramount.

In April 1989 the United States Patent and Trademark Office *granted* the first patent on an animal to the President and Fellows of Harvard College. Researchers[22] there had produced transgenic mice whose germ and somatic cells contained an activated onco-gene sequence. The sequence was introduced into the animal at its embryonic stage. The result of the genetic modification was an increase of the probability of the development of malignant tumors.

Two applications for a patent were made to the EPO and triggered, *inter alia*, an invocation of Article 53(a) EPC.[23] The first application related to a process claim and concerned a method of producing a genetically engineered mouse. The second application related to a product claim and was, essentially, a claim to the mouse itself.

In contrast to the United States Patent and Trademark Office, the Examining Division of the EPO rejected the applications initially.[24] During the hearing the Examining Division argued that patent law, and Article 53(a) in particular, was not the correct legislative tool for *regulating* problems arising in connection with genetic modification of organisms.[25]

---

20   See Comments on the first Preliminary Draft Convention relating to a European patent law, meeting held at Bonn 14 Mar 1961, Document IV/2071/61-E, available from the EPO.

21   [1991] EPOR 525. In 2003 an interlocutory decision was issued by the Opposition Division of the EPO which was appealed. On 6 July 2004 the Technical Board of Appeal further restricted the scope of the patent to cover only 'transgenic mice'.

22   Philip Leder and Timothy Stewart.

23   Decision of Technical Board of Appeal 3.3.2 dated Oct 1990. T 19/90 3.3.2.

24   July 1989.

25   See the Text of the Decision and the arguments of the Examining Division entitled 'Considerations under Art. 53(a) EPC' in the Reasons for the Decision at point 5.

On appeal, the Technical Board disagreed with the Examining Division. The Board said that in a case like the present there were compelling reasons why the implications of Article 53(a) EPC should be considered. First, there was the fact that insertion of onco-genes into mammalian animals could cause suffering. Second, there was a danger that genetically modified organisms when released into the general environment could cause irreversible adverse effects. According to the Board, the decision on the applicability of Article 53(a) depended upon weighing the suffering of animals and possible risks to the environment on the one hand, and the invention's usefulness to mankind on the other. On this basis, the Board remitted the applications to the Examining Division for further prosecution, whereupon the Examining Division decided in the applicant's favour. One of the issues for determination was whether or not the subject-matter of the applications was contrary to *ordre public* or morality within the meaning of Article 53(a) EPC.[26]

In view of the decision of the Board, and the importance the public attached to patenting animals, the Examining Division considered it appropriate to issue a statement on the matter. The text of the statement[27] contains what may be described as EPO 'policy' on patenting transgenic animals. The EPO said that, in development, all new technology is accompanied by risks, which itself should not lead to a negative attitude towards the technology in question. Rather, the potential risks of the technology and its beneficial results should be carefully weighed, so that an informed opinion can be reached as to whether or not the technology should be used.[28] In respect of patenting a particular invention morality must be examined, and possible detrimental effects and risks must be weighed and balanced against the advantages aimed at. On this basis, in the present case there were three different interests to be considered, namely:

- those of mankind in remedying dangerous diseases;
- protection of the environment;
- those of animals suffering in testing.

The Examining Division considered that protection of the environment, combined with suffering caused to animals, might well preclude the invention from patentability by virtue of Article 53(a), *unless* the benefit to mankind could outweigh such negative concerns. The Examining Division proceeded to assess the benefit to mankind and invoked the following considerations, which, it said, particularly applied:

---

26   The other issue was whether or not the subject-matter constituted a variety within the meaning of EPC Art. 53(b).

27   See Annex to Form 2035.3 /2004, entitled 'Comments'.

28   See the Text of the Decision at point 3.

- Since cancer was one of the most frequent causes of premature death in many countries of the world, the development of new and improved anti-cancer treatments must be regarded as valuable. Hence, the usefulness of the invention to mankind could not be denied.
- The induction of cancer in one animal meant that the numbers of animals required for conventional testing was reduced. Hence, the invention resulted in a reduction of the overall number of animals used for experimental purposes.
- With regard to whether or not alternatives to animal testing existed in the particular context, studies[29] showed that in cancer research at the time animal test models were indispensable.
- In respect of possible risks to the environment the purpose of the invention must be considered. Since the purpose of the invention was to permit the use of animal test models exclusively in a laboratory, under controlled conditions by qualified staff, no deliberate release into the environment was intended. Therefore, the risk for uncontrolled release lay either with an intentional misuse of the technology, or blatant ignorance on the part of staff carrying out the tests. The mere fact that such uncontrollable acts were conceivable could not be a major determinant in deciding whether or not to grant patent.
- Exclusion from patentability could not be justified merely because a technology was dangerous. Regulation for handling dangerous material was not the task of the EPO but of specialized government authorities.

Taking these factors into consideration, the Examining Division concluded that the invention could not be considered immoral so as to preclude it from patentability by virtue of Article 53(a). However, the Examining Division stressed that the considerations it outlined applied *solely* to the present case, and other instances of transgenic animals might arise in the future for which a different outcome could be reached in applying Article 53(a).

The Examining Division statement is worthy of several comments:

- Development of all new technology is accompanied by risks, necessarily including biotechnology. The implication, therefore, is that, for the purposes of patent law, biotechnology must be treated in the same manner as any other technology. In this respect the EPO adopted a sensible position. To raise substantially the threshold of a morality test would include categories of inventions, which, until now, have been considered outside the range of the morality exclusion. Examples of such categories are inventions relating to nuclear energy, anti-personnel devices, chemical warfare agents, defoliants and battery cultivation.

---

29   See generally A Berns, 1991 vol. 1 *Current Biology* at 28.

- All types of technology require assessment under Article 53(a) EPC. The implication is that biotechnological inventions in general, and genetic engineering inventions in particular, are not per se excluded from patent protection.
- Threats to the environment are within the ambit of the morality provision. However, Article 53(a) prohibits patenting of inventions the publication or exploitation of which would be contrary to *ordre public* or morality. Environmental threats are either in respect of consequence, or impact, upon the environment, or in relation to dominion of man over the environment. EPO policy suggests that environmental threats fall under the heading of 'morality' only, and not '*ordre public*'. However, it is possible that environmental impact considerations might fall under the heading '*ordre public*'.[30] In *Onco-Mouse* the Examining Division missed an opportunity to elaborate upon the meaning of environmental threat in the context of *ordre public* and to clarify the law in this area.
- The mere fact that uncontrollable risks are conceivable cannot be a major factor in determining whether or not a patent can be granted. In the absence of intention, it may be difficult to sustain a complaint of 'threat to the environment'; if so, a patent application can rarely expect to be refused on the basis of environmental threat alone.
- The EPO laid down criteria for determining when Article 53(a) applies, namely, a balancing of interests between the suffering caused to animals and risks posed to the environment on the one hand, and beneficial effects to mankind on the other. However, there was no suggestion in *Onco-Mouse* of any method for assessing or weighing such criteria. Criteria without proper guidance for application are of little benefit to patent applicants.
- The EPO stressed that the considerations it outlined related *solely* to the *Onco-Mouse* case. It is difficult to argue forcibly that any EPO 'policy' has in fact been mooted in respect of the application of Article 53(a) EPC.
- Insisting all technology be assessed under Article 53(a) is extending EPO jurisdiction beyond that outlined in the Guidelines for Examination. The Guidelines suggest a 'fair test' approach should be adopted in relation to morality. Such a test requires an overwhelming consensus that exploitation of the invention would be immoral[31] before patentability can be excluded. However, according to the EPO the application of Article 53(a) is based on a 'balancing of interests' which is wider than that envisaged under the 'fair test' approach.

Despite this criticism, the decision in *Onco-Mouse* is welcome. The author does not see the general public expressing repugnance towards the invention, nor animal loss as out of proportion to the potential human benefit. However, the later case of *Upjohn's*

---

30   See the *PGS* case [1995] EPOR 357, discussed hereafter.
31   See *Howard Florey/Relaxin* [1995] EPOR 541, discussed hereafter.

Application[32] is borderline. In this case the invention related to a transgenic mouse modified to provide a model to accommodate research into methods for stimulating hair growth. The main objection of the examiners related to lack of inventive step as regards the invention as a whole. However, there was an additional objection on moral grounds. This related to the fact that one of a number of alternative genes suggested for insertion into the mouse was an onco-gene. Because it was possible to delete that single gene without detriment to the breadth of the main claim, Upjohn were able to redraft their application. The remaining claims were held to be valid and the patent allowed. Whether or not the additional objection would have been sustained on appeal is conjecture and will never be known.

However, the case involving the University of Edinburgh (the so-called Edinburgh patent) is more instructive.[33] The University of Edinburgh sought to patent animal transgenic stem cells. The EPO granted the patent after the introduction of the Biotechnology Directive. Due to international reaction against the patent, and oppositions by the governments of Germany, Italy and the Netherlands, the University amended the claims limiting the application to non-humans. The Opposition Division finally held that, while the original application violated Article 53(a) EPC, the amended claims did not.[34] This suggests that the EPO is reluctant to apply the morality provision outlined in the EPC and the Directive.

Human Genetics and Article 53(a)

In *Howard Florey/Relaxin*[35] (henceforth *Relaxin*), the application of Article 53(a) was again in issue. The case concerned a patent in respect of a genetically engineered human hormone that had no previously recognized existence.[36] The patent was granted by the Examining Division of the EPO in April 1991 and later opposed. The opponents maintained that the claimed DNA sequence offended against Article 53(a) EPC. The Opposition Division rejected the objections on all grounds. One of the opponents lodged an appeal which is currently pending before the Technical Board of Appeal.[37]

In the Reasons for the Decision, the Opposition Division discussed the application of Article 53(a). It said:

---

32   Case no. 89913146.0. See generally Robin Nott, 'Plants and Animals: Why They Should Be Protected by Patents and Variety Rights', 1993 July/Aug *Patent World* 45 at 47.

33   European Patent No. 695 351, issued 8 Dec 1999.

34   Press Release, EPO, 'Edinburgh Patent Limited After EPO Opposition Hearing', 24 July 2002.

35   [1995] EPOR 541.

36   Discussed again more fully in Ch. 7 hereafter in the context of patenting genes and the new Directive on the Legal Protection of Biotechnological Inventions 1998.

37   There has been no ruling by the Technical Board to date.

The function of the Article has to be seen as a measure to ensure that patents would not be granted for inventions universally regarded as outrageous. This interpretation reflects the 'fair test' provision in the relevant passages of the Guidelines for Examination,[38] which state that Article 53(a) is likely to be invoked only in rare and exceptional cases.[39]

By endorsing the test outlined in the Guidelines, the Opposition Division is supporting a 'light' moral regime in respect of the application of Article 53(a). Article 53(a) constitutes an exception to the general principles of patentability, set out in Article 52(1) EPC, namely, that patents shall be granted for inventions, in all fields of technology,[40] provided they are industrially applicable, novel and inventive.[41] Boards of Appeal have repeatedly construed such exceptions narrowly.[42] According to the Opposition Division, it would be abhorrent to the overwhelming majority of the public if the invention involved the patenting of human life, an abuse of pregnant women, or a return to slavery and the piecemeal sale of women. However, the Opposition Division denied that the invention was comparable to any of these things.[43]

The Opposition Division decision provides guidance for future cases as to what constitutes an abhorrent invention precluded from patentability. In relation to assertions concerning the alleged intrinsic immorality of patenting human genes, the Opposition Division said that these were founded on the premise of an overwhelming consensus among Contracting States that the patenting of human genes is abhorrent and hence prohibited under Article 53(a).[44] Such an assumption was false, in its view.

In relation to the application of Article 53(a), the 'overwhelming consensus' criterion adopted by the Opposition Division in *Relaxin* is more sensible than the 'balance of interests' test adopted by the Examining Division in *Onco-Mouse*. The 'overwhelming consensus' test limits refusal of patentability by the EPO. The decision supports the proposition that consensus among Contracting States is the controlling factor. There is no inherent power in the EPO or other European institutions to decide what constitutes an abhorrent invention. The decision suggests that the intention underlying the drafting of Article 53(a) was to continue states' practice in respect of power to refuse patentability on the basis of morality.[45]

---

38   Guidelines for Examination of the European Patent Office, *supra*.

39   Reasons for the Decision at 6.2.1; and the Guidelines for Examination, *supra*.

40   As amended by EPC 2000.

41   Ibid. at 6.2.2.

42   See *Lubrizol/Hybrid Plants* [1990] EPOR 337 at point 6 of the Reasons for the Decision; and *Harvard/Onco–Mouse* [1990] EPOR 501 at point 3 of the Text of the Decision.

43   Reasons for the Decision at point 6.3 and discussed more fully in Ch. 7 hereafter.

44   Ibid. at point 6.4.3.

45   See generally Ch. 2 heretofore.

Under the EPC there was no intention to permit the EPO to concern itself with moral issues in a wider direction. As the Opposition Division is part of the EPO and its views appear to be so different from those of the Examining Division, confusion clearly exists in the Office in regard to the application of Article 53(a).

In a trilogy of more recent cases involving breast cancer genes, the application of Article 53 (a) was again in issue. All three cases involved the University of Utah Research Foundation (as patent proprietor) and Opposition Division decisions. The cases were referred to the Technical Board of Appeals. Case *T 1213/05*[46] involved an appeal by the patent proprietor (and opponents[47]) against a decision of the Opposition Division according to which European Patent No. 0 705 902 could be maintained in amended form (the Opposition Division had decided that the main request did not meet the requirements of Article 123(2) and (3) and Article 84 EPC). The patent was entitled '17 q-linked breast and ovarian cancer susceptibility gene'.

In considering the appeal, the Technical Board heard arguments, *inter alia*, as to why the claimed subject-matter was excluded from patentability under Article 53(a). In the first place, it was argued (by opponents) that no proof existed that donors of the cells that had been critical to identify the breast cancer gene (BCRA1) had given a previous informed consent. In the absence of such proof, it had to be assumed that the initial obtaining of the research results involved severe ethical violations, and thus a violation of *ordre public* or morality as referred to in Article 53(a). The Board responded by saying that the EPC contains no provision establishing a requirement for an applicant to submit evidence of a previous informed consent. Accordingly, the Board did not accept that the claimed subject-matter offended against Article 53(a). The second argument by the opponents was that the socio-economic consequences of the patenting of the claimed subject-matter should be considered by the Board because these consequences touched ethical issues in the sense that the claimed subject-matter would not only result in increased costs for patients, but would also influence the way in which diagnosis and research would be organized in Europe. In response, the Board looked to the pertinent wording of Article 53(a) and noted that it forbade patenting where 'exploitation of the invention' and not 'exploitation of the patent' would be contrary to *ordre public* or morality. On this basis, the Technical Board rejected this objection.

In case *T 0666/05*[48] appeals were again lodged by the patent proprietor (and opponents) against the decision of the Opposition Division according to which European Patent no. 0 705 903 could be maintained in amended form. The patent was entitled 'mutations in the 17 q-linked breast and ovarian cancer susceptibility gene'. The opponents argued that methods for diagnosing a predisposition for breast and ovarian cancer should not be regarded as patentable inventions according to Article 53(c), which forbids patenting of inventions in respect of methods for

46   Decision of the Technical Board of Appeal, 27 Sept 2007.
47   *Greenpeace e. V. et al.*
48   Decision of the Technical Board of Appeal 13 Nov 2008.

treatment of the human or animal body by surgery or therapy, and diagnostic methods practiced on the human or animal body. The Board's response was that Article 53(c) excludes diagnostic methods practiced on the human or animal body only if the method steps of technical nature belonging to the preceding steps which are constitutive for making a diagnosis are performed on a living human or animal body.[49] The opponent's argument was, therefore, rejected on this ground. The opponent also argued that the claimed subject-matter contravened the requirements of Article 53(a), in that it involved ethical issues. The Technical Board also rejected this argument, preferring to follow its decision in case *T 1213/05*.

Again in case *T 0080/05*, the Technical Board of Appeal considered the application of Article 53(a).[50] The appeal was lodged by the patent proprietor against the decision of the Opposition Division according to which European Patent No. 0 699 754 was revoked. The Technical Board set aside the decision of the Opposition Division. The Board rejected arguments based on Article 53(a), that if the patent was granted patients would no longer be able to have their genetic information read and interpreted by the organization of their choice; and it could not be guaranteed that criminal and medical gene databases were kept strictly separate, which was an accepted ethical principle of the EPO. The Board again preferred to follow its decision in case *T 1213/05* and, on this basis, rejected arguments based on ethical considerations.

What the cases demonstrate is the need for a uniform approach by the Examining Division, the Opposition Division and the Technical Board of Appeal, all part of the EPO. A policy on the extent, meaning and interpretation of Article 53(a) is required.

Plant Genetics and Article 53(a)

With respect to plants, the issue of morality first arose in the *Lubrizol Transgene Expression* case, 31 March 1992.[51] Here, several non-governmental organizations filed notices of opposition in which they claimed patents on plants were immoral by virtue of Article 53(a). The invention related to the creation of new plants whose nutritive value exceeded that of conventionally obtained plants. The Opposition Division decided that the exclusion from patentability outlined in Article 53(a) dealt with extreme cases universally regarded as abhorrent. Hence, considering that the plants covered by the patent might give rise to a better management of food

---

49   Quoting from the Enlarged Board of Appeal in its Opinion G 1/04 (OJ EPO 2006, 334).

50   Decision of the Technical Board of Appeal 19 Nov 2008.

51   Opposition Division, 31 Mar 1992, concerning European patent application bearing publication no. 122.791; see Geertrui van Overwalle, 'Biotechnology Patents in Europe: From Law to Ethics', in Sigrid Sterckx (ed.), *Biotechnology, Patents and Morality* (1997) 138–48 at 141, Ashgate Publishing, UK.

shortage in the world, the Opposition Division ruled that there was no violation of Article 53(a).

The case is important for three reasons. First, it was the first case in which the application of Article 53(a) was considered in respect of plants. Second, it established that plants per se are not excluded from patentability. Third, it established that application of Article 53(a) is limited to inventions universally considered abhorrent. This again supports the proposition that a 'light' moral regime is effected by the Article.

In *Plant Genetic Systems N.V. et al.*[52] (henceforth *PGS*), the application of Article 53(a) was again in issue. The patent concerned genetically modified plants and plant cells, made resistant to glutamine synthetase inhibitors (henceforth GSI) by the insertion of a single novel gene. The specification disclosed the use of modern biotechnological techniques for the production of GSI-resistant plants and seeds, which contained heterologous DNA encoding a protein capable of inactivating the GSI herbicides. In this manner, a new trait was added to the genetic material of cells and allowed the plant to grow even in the presence of the inhibitor. Greenpeace Ltd, the environmental action group, opposed the patent. The Opposition Division rejected the opposition.[53] Greenpeace then appealed to the Technical Board of Appeal.

One of the main issues in the appeal was whether or not any of the claimed subject-matter constituted an exception to patentability by virtue of Article 53(a) EPC. The Board looked to the historical documentation of the Patents Working Party in relation to the creation of Article 53 EPC.[54] The documents recognize that since there is no European concept of *ordre public* or morality, the interpretation of these concepts is a matter for European institutions.[55] The Board in *PGS* concluded there was a general acceptance[56] that the concept of *ordre public* covered *public security*, *protection of the environment* and the *physical integrity* of individuals.[57] Finally, for the Board, the concept of morality was concerned with the belief that some behaviour was right and acceptable, whereas other behaviour was wrong. Within the context of the EPC, this belief was rooted in the culture inherent in European society and civilization.[58] The Board rejected arguments of the appellant based on surveys and opinion polls because they were not probative of public opinion, and did not necessarily reflect concerns of *ordre public* or morality.[59] The Board justified rejecting surveys and opinion polls by analogy with the provision

---

52   [1995] EPOR 357.

53   Decision of 15 Feb 1993.

54   Minutes of the Proceedings of the first meeting of the Patents Working Party held at Brussels from 17 to 21 Apr 1961, Document IV/2767/61-E, available from the EPO.

55   Ibid. Minutes of the meeting held on 18 Apr 1961.

56   Presumably by states.

57   See Reasons for the Decision at point 5.

58   Ibid. at point 6.

59   Ibid. at point 15.

in Article 53(a) second half-sentence: just as national laws and regulations either approving or disapproving of the exploitation of an invention are not decisive on *ordre public* or morality, neither are (national) surveys or opinion polls.

Three points can be made. First, the Board accepted that interpretation of Article 53(a) was a matter for European institutions. EPO jurisdiction to refuse patentability on the basis of morality was justified by the historical documentation of the EPC. However, historical documentation reveals[60] an absence of intention on the part of states to permit the EPO authority to explore moral issues in a wider direction. Therefore, the jurisdiction of the EPO in respect of previously unknown moral issues is questionable. Second, jurisdiction of the EPO was expanded on the basis of an alleged consensus among states. It is uncertain whether or not there is *acceptance* among states that the concept of *ordre public* relates to public security, protection of the environment and the physical integrity of individuals. In enunciating such a broad proposition the EPO conferred upon itself a role corresponding to that of a 'supranational' authority in EU law. In this manner the EPO established itself as the de facto arbiter of *ordre public* concerns among states, whereas it has no de jure jurisdictional basis. Third, according to the Board, the concept of morality is related to a 'belief' rooted in the *culture inherent* in European society and civilization. However, there is scant evidence suggesting a culture inherent in European society. If there were such a culture, Article 53(a) second half-sentence, namely, that relating to law and regulation in some or all of the Contracting States, would be redundant. This part of the provision is an acknowledgement that behaviour acceptable in some states can be unacceptable in others.

Other arguments of the appellants suggested that exploitation of the invention would damage the environment. As noted hereafter, the Board considered this objection involved both morality and *ordre public* concerns. It related to the former in respect of the dominion gained by man over the natural world and to the latter in regard to alleged environmental consequences.

Morality and environmental concerns   The appellant's argument was based on the assumption that exploitation of the invention enabled man to control the natural world, which, in their view, was contrary to morality. The Board disagreed. The basis of its decision lay in refusing to accept that a difference exists between traditional selective breeding, and breeding as a result of a biotechnological process. The Board said that plant biotechnology per se should not be regarded as more contrary to morality than traditional selective breeding. Both traditional breeders and molecular biologists are guided by the same motivation, namely, to change the property of a plant by introducing novel genetic material into it, in order to obtain a new (and possibly improved) plant.[61] According to the Board, it was important to understand the basis of genetic engineering and how it worked. Genetic engineering was fundamentally analogous to selective breeding in that

---

60   See generally Ch. 2 heretofore.
61   See Reasons for the Decision at point 17.1.

the same result was achieved, albeit much more quickly. However, to see genetic engineering merely as an expedited process was not enough. The significance of genetic engineering lay in its capacity to innovate, which in the present case meant the ability to produce transgenic plants. Protecting innovation by means of patent law was legitimate, even for the more mundane function of expediting breeding processes.[62]

Ordre public *and environmental concerns*   The appellant's argument here was that patentability should be denied on the basis that exploitation of the invention would seriously prejudice the environment. The Board again disagreed. Insufficient evidence was adduced to satisfy the 'balance of interests' test outlined in *Onco-Mouse*. The Board said that, while patent offices stood at the crossroads between science and public policy, potential risks in relation to any given invention are not adequately assessed merely from disclosure in the patent application. The right to exploit an invention was not unconditional, but was determined within the legal framework defined by national laws and regulations regarding its use.[63] Assessment of the hazards stemming from exploitation of a given technology was one of the more important functions of *regulatory authorities*, which are in a position to carry out a realistic evaluation of risks on the basis of regulations in force, objective criteria and scientifically valid parameters.[64] The Board concluded that to revoke a patent on the grounds that exploitation would seriously damage the environment presupposed that the threat was sufficiently substantiated at the time the decision to revoke the patent was taken by the EPO. As the appellants had merely shown that some damage to the ecosystem might occur, assessment of patentability on the basis of the so-called 'balancing exercise' of benefits and disadvantages was inappropriate. However, despite reservations surrounding the 'balance' test, the Board was of the opinion that it is undoubtedly contrary to *ordre public* or morality to propose either a misuse, or a destructive use, of the techniques involved.[65] Accordingly, the Board looked to the aim of the invention, which, it said, was essentially to produce seeds and plants resistant to a particular class of herbicide. The Board concluded that none of the claims referred to subject-matter which related to either a misuse, or a destructive use, of plant biotechnological techniques.

To summarize, *Onco-Mouse* suggests, unless clear evaluation criteria emerge in respect of *ordre public* and morality, the EPO is unable to offer sufficient guidance for future cases as to the meaning and application of Article 53(a). This supports the view that the ability of patent examiners to rule on matters moral is questionable. *Howard Florey/Rexaxin* suggests in the application of Article 53(a) confusion

---

62   This line of reasoning also applies to animals according to Barry Hoffmaster, 'The Ethics of Patenting Higher Life Forms', 1988 vol. 4 *Intellectual Property Journal* 1–24 at 16.

63   See Reasons for the Decision at point 18.2.

64   Ibid. at point 18.4.

65   Ibid. at point 17.1.

clearly exists within the Office. This supports the view that the appropriateness of including morality provisions in patent law at all is problematic. *PGS* suggests the acceptability of tests laid down by the Office is suspect. This supports the view that what is required is a definition of *ordre public* and morality concerns based on states' practice. What the jurisprudence does show is that Article 53(a) as a mechanism for determining exceptions to patentability is questionable.

Application of Article 53(b)

The application of Article 53(b) by the EPO is examined in this section. The jurisprudence is unclear and the cases demonstrate a conflict between decisions of Technical Boards of Appeal in regard to the scope of protection and the extent of exclusion of certain subject-matter from patentability. This is likely to result in inconsistency of practice among EPC states, and supports the view that Article 53(b) as a mechanism for determining exclusions from patentability is questionable.

As already noted, Article 53(b) EPC deals with subject-matter exclusion and reads as follows:

> European patents shall not be granted in respect of:
>
> (b) plant or animal varieties or essentially biological processes for the production
> of plants or animals; this provision shall not apply to microbiological processes
> or the products thereof.

Difficulties with this provision include that Article 53(b) represents an exception to the general provision of Article 52(1) EPC, according to which European patents shall be granted for inventions, in all fields of technology, which are susceptible of industrial application, new, and involve an inventive step. Further difficulties arise by virtue of the revised UPOV Convention (Act of 1991), according to which certain plant varieties are not eligible for protection thereunder. The result is that protection of life forms by the patent system is limited.[66] Arguably, however, most problematic for breeders is the manner in which the provision is drafted. The cases discussed hereafter testify that breeders cannot rely on the uncertain protection offered by complex legislation. In respect of patentability, legislation drafted in clear, positive terms is required. In particular, what the provision means by the terms 'variety', 'essentially biological' and 'microbiological processes' needs clarification.[67]

---

66   Except for micro-organisms, which have long since enjoyed patent protection.
67   In this regard, the Patents Act 1977 UK is amended by the Patents Regulations 2000.

According to *Webster's Dictionary*,[68] a variety is:

> a group of animals or plants related by descent but distinguished from other similar groups only by characters considered too inconsistent or too trivial to entitle it to recognition as a species, or whose distinguishing characters are dependent on breeding controlled by man for their perpetuation; often, any group of any rank lower than a species.[69]

The term 'variety' is not defined in the EPC, and there is no generally defined taxonomic definition as there is for 'species' or 'genus'. Such a situation leads to difficulties in legal interpretation.

Plant Genetics and Article 53(b)

In *Ciba-Geigy/Propagating Material*[70] (henceforth *Ciba-Geigy*), the term 'variety' was interpreted to mean 'a multiplicity of plants which are largely the same in their characteristics and remain the same within specific tolerances after every propagation or every propagation cycle'. Here, the invention related to the treatment of plant propagating material by a sulphurous oxime derivative so as to protect it from herbicides. The Examining Division rejected the application on the grounds that what was claimed offended against Article 53(b) EPC, namely, that the subject-matter amounted to a plant variety. The Board of Appeal disagreed. It held that, since treatment of the propagating material with the oxime derivative could also be carried out on propagating material not meeting the essential criteria of a plant variety, the subject-matter was patentable.[71] The Board said: 'Article 53(b) EPC prohibits only the patenting of plants or their propagating material in the genetically fixed form of the plant variety'.[72]

Three comments may be made. First, in limiting the exclusion to the genetically fixed form of the plant variety the EPO interpreted the sub-section in a restrictive manner. This approach accords with EPO jurisprudence in respect of Article 53(a) and is welcome. Patentability was permitted on the basis that the invention was generic, in the sense that it applied to more than a single plant variety. However, the decision suggests that not only is a variety per se excluded from patentability, but that individual plants within the variety so created are also precluded. This indicates that patent applications with claims directed to a 'sub-species' will be considered as falling within the ambit of the exclusion, and, hence, not patentable. Second, the characteristics referred to by the Board are those of 'homogeneity'

---

68 *Webster's New International Dictionary of the English Language*, 2nd edn unabridged (1952), 2819 Springfield, Massachusetts, Merriam Publishing.

69 See Rainer Moufang, 'Patentability of Genetic Inventions in Animals', *supra* at 832.

70 [1979/85] EPOR vol. C.

71 In other words, the invention was not limited to plant varieties only.

72 See Reasons for the Decision at point 2.

and 'stability'. Homogeneity refers to the fact that the variety must be sufficiently uniform in its relevant characteristics.[73] Stability refers to the fact that the variety must remain unchanged after repeated propagation.[74] While the decision in the case is welcome, it is regrettable that the Board felt it necessary to look to the UPOV system for a definition of the term 'variety'. Patent protection was possible only because the claimed subject-matter fell outside the UPOV definition of 'variety'. The *logic* of the decision suggests that varieties unprotectable under the 1991 Act of UPOV should be eligible for patent protection. In this sense the decision is also welcome.[75] However, patentability should be determined within the parameters of patent law, without reference to other systems of protection. Third, the Board recognized that the invention claimed did not lie within the sphere of plant breeding, as it was not concerned with genetic modification of plants. The innovation lay in the treatment of seed by chemical agents for the purpose of rendering it resistant to agricultural chemicals. The invention was generic and related to more than a single variety. The invention was patentable. Such a broad view of invention is welcome.

In *Lubrizol/Hybrid Plants*[76] (henceforth *Lubrizol*), the question of what amounted to a 'variety' was again in issue. Here, the invention related to the use of a heterozygous parent plant for propagation purposes ensuring development of selected, and desired, hybrid plants and seed in a repeatable manner. In heterozygous plants both chromosomes carry alternative forms of the same gene for a given trait. In homozygous plants, on the other hand, both chromosomes carry the same gene. Only when homozygosity is provided can a certain generation be produced repeatedly as a crossing result. In other words, only homozygous plants can breed 'true'.

The invention was said to reside in the use of a heterozygous parent plant to produce hybrids with selected characteristics *as if* that parent were of a homozygous nature. The claims related to processes for rapidly developing hybrids and for hybrid seed in general. The Examining Division rejected the application on the basis that the claims offended against Article 53(b) EPC. The Technical Board of Appeal disagreed. The Board said the term 'variety' encompassed the twin characteristics of homogeneity and stability. According to the Board, these characteristics were a prerequisite constituting a plant variety. The Board considered whether or not the claimed products themselves amounted to a variety so as to be unprotectable by virtue of Article 53(b). The Board concluded that the claimed hybrid seed and plants were not 'stable' when considered as a whole generation population; they did not, therefore, amount to a variety and consequently were patentable.

---

73   1961 Act Art. 6(c) and 1991 Act Art. 8 UPOV.

74   1961 Act Art. 6(d) and 1991 Act Art. 9 UPOV.

75   However, in the light of the decision in *PGS* [1995] EPOR 357, this is unlikely to occur.

76   [1990] EPOR 173.

While the decision of the Board is welcome, it is not logical. The Board considered that, because hybrid plants result from a mix of parents, they are never stable. Since the 'twin characteristics' criterion was not complied with, the subject-matter of the claims did not amount to a variety so as to preclude patentability. The Board expressed the view that, even if the hybrid generation was comprised of single individual plants which could be stable for a certain trait when further crossed and propagated, this fact did not contradict the stated non-stability of the generation as a whole.[77] According to the Board, single individual plants were not embraced within the subject-matter of the product claim.[78] The logic of the Board is difficult to sustain because there is a conflict between the stated non-stability of the generation as a whole, with the (stated) stability of individual plants comprising that generation. If individual plants do not constitute a 'variety', what does?

Thus, while there was agreement between the Boards in *Ciba-Geigy* and *Lubrizol*, namely, that, the term 'variety' encompassed the 'twin characteristics' criterion, there was no consensus as to the application of Article 53(b).

In *Plant Genetic Systems N.V. et al.*,[79] the meaning of Article 53(b) was considered yet again. Here, plant cells were made resistant to certain herbicides by genetic modification through the insertion of a single novel gene. The claims related to processes and products of the herbicide 'Basta'.[80] One of the issues before the Board of Appeal was whether or not the claimed subject-matter constituted an exception to patentability by virtue of Article 53(b). The first consideration, therefore, was to determine whether the claimed subject-matter amounted to a 'variety'.

The Board reviewed existing EPO jurisprudence. It noted that in *Ciba-Geigy* what was precluded from patentability was the patenting of plants or propagating material in the genetically fixed form of the plant variety. Since, in that case, treatment of the propagating material could also be carried out on plants not meeting the essential criteria of a plant variety, the invention was not limited to one variety only. Consequently the invention was patentable. The Board also noted that since the claims in *Lubrizol* related to hybrid material, the resulting seed or plants could not be stable. Hence, the invention in *Lubrizol* was not directed to a plant variety as understood within the context of the UPOV Convention.

Prima facie, therefore, the Board in *PGS* endorsed the views of other Technical Boards as to the meaning of the term 'variety'. It seemed as if EPO 'policy' was emerging to offer guidance for future cases. Unfortunately this was not the case, since the Board in PGS proceeded to hold the invention not patentable on the grounds that the claimed subject-matter did amount to a plant variety after all. It said:

---

77 See Reasons for the Decision at point 14.

78 Here the Board is differentiating between a 'variety' and individual plants comprising that variety.

79 [1995] EPOR 357.

80 See Margaret Llewelyn, 'Article 53 Revisited', 1995 vol. 10 *European Intellectual Property Review* 506–11 at 506.

> In the Board's judgment, the concept of 'plant varieties' … [r]efers to any plant grouping … [w]hich … [i]rrespective of whether it would be eligible for protection under the UPOV Convention, is characterised by at least *one single* transmissible characteristic distinguishing it from other plant groupings and which is sufficiently homogeneous and stable in its relevant characteristics.[81]

The Board's approach has been criticized by Tim Roberts,[82] who argues correctly, it is submitted, that such a statement shows a profound misunderstanding of the nature of a UPOV plant variety. Under UPOV, a plant variety is not characterized by a single stable gene, but rather by essentially all of its genes, or at least those genes expressing and determining plant phenotype. It is submitted a plant grouping characterized by a single novel gene is a generic invention and not a plant variety.

The decision of the Board in *PGS* is suspect for two reasons. First, it runs counter to the principle enunciated in *Ciba-Geigy*, namely, that it is immaterial to the question of patentability whether the propagating material could also be, or is primarily, a plant variety. Second, the decision of the Board also runs counter to the principle outlined in *Lubrizol*, namely, that single individual plants should not be construed as embraced within the subject-matter of the product claim. The decision of the Board in *PGS* is in conflict with decisions of earlier Boards in regard to the scope of Article 53(b) and the extent of exclusion from patentability. There is no uniform approach by the EPO to interpreting Article 53(b).

More recently, in *Novartis/Transgenic plant*[83] (henceforth *Novartis*), the Technical Board of Appeal addressed the issue of whether or not the exclusion of plant varieties within Article 53(b) extended to plant groupings which comprised a number of varieties. The case related to an application for transgenic plants and processes for creating them. The Board rejected the argument that the exclusion only related to plant varieties as such. The Board held that Article 53(b) was intended to apply not merely to claims for a plant *variety* but claims for plant *varieties*, and the exclusion extends to claims for plant groupings which comprise more than one variety even if a single variety is not being claimed.[84] The Board failed to address the question of whether its concept of a claim which encompasses a plant variety was to be broadly applied to cover any claim to plant material on the basis that it could, potentially, relate to a variety suggesting that that the breadth of the exclusion remains unclear for breeders; it is, therefore, uncertain as to what plant material can safely be regarded as patentable.[85] On request from Novartis,

---

81   See Reasons for the Decision at point 23.

82   See Tim Roberts, 'Patenting Plants Around the World', 1996 vol. 10 *European Intellectual Property Review* 531–6 at 535.

83   [1999] EPOR 123.

84   See Margaret Llewelyn, 'European Plant Variety Protection: A Reactionary Time', 1998/1999 vol. 6 *Bio-Science Law Review* 211–19 at 213.

85   Ibid. at 214.

the Technical Board referred, *inter alia*, the following point of law to the Enlarged Board of Appeal: 'Does a claim which relates to plants but wherein specific plant varieties are not individually claimed *ipso facto* avoid the prohibition on patenting in Article 53(b) EPC even though it embraces plant varieties?'

The Enlarged Board of Appeal[86] has since upheld Novartis's argument that a claim encompassing more than one variety is not excluded under Article 53(b) EPC, suggesting that the comments of the Technical Board are effectively redundant. The EBA held that Article 53(b) did not exclude anything other than plant varieties which can be protected under plant variety rights. The result of the decision appears clear: in order for the exclusion of plant varieties to apply the claim must be directed to a variety capable of protection by a plant variety right.[87]

Although the EBA decision in *Novartis* reflects a wider pro-patenting attitude in respect of protecting biotechnological inventions, the law in respect of plant genetics and the application of Article 53(b) remains unclear for breeders.[88] What is required is clear EPO policy on interpretation.

Animal Genetics and Article 53(b)

Difficulty as to the meaning of the term 'variety' also exists in relation to animal genetics. In *Harvard Onco-Mouse*[89] the question arose as to what amounted to an 'animal variety'. The main claims related to a method of producing a transgenic mouse and to the mouse itself. The Examining Division ruled that Article 53(b) first half-sentence, by rendering animal varieties unpatentable, meant that animals per se were also excluded from patentability. On this basis, the transgenic mouse was unpatentable. The Examining Division also ruled that, even if the transgenic mouse was the result of a microbiological process under Article 53(b) second half-sentence, this did not render it patentable. The Board of Appeal disagreed, and remitted the application for further consideration by the Examining Division in two respects,[90] namely:

1. In interpreting what 'animal variety' means, the Examining Division must find the common meaning of this term in the three official languages[91] of

---

86   Decision of 20 Dec 1999 at point 3.10 of Reasons for the Decision.

87   See Margaret Llewelyn, 'The Legal Protection of Biological Material in the New Millennium: The Dawn of a New Era or 21st-century Blues?', 1999/2000 vol. 4 *Bio-Science Law Review* 123–30 at 124.

88   There are some references to the Enlarged Board of Appeal that are currently pending on plants. Case G 2/07 regarding a method for selective increase of the anticarcinogenic glucosinolates in brassica, and Case G 1/08 regarding a method for breeding tomatoes having reduced water content and product of the method. These will be considered in consolidated proceedings.

89   [1991] EPOR 525.

90   See Reasons for the Decision at point 4.8.

91   English, French and German.

the EPC. If the Division comes to the conclusion that the subject-matter is not covered by any of these three terms, then Article 53(b) EPC constitutes no bar to patentability.

2. The Examining Division must consider, should the case arise, whether the claimed processes constitute microbiological processes within the meaning of Article 53(b) EPC.

On re-examination, the Examining Division ruled that Article 53(b) first half-sentence did not preclude patentability of the transgenic mouse. It said:

> [a]lthough the term animal variety is not entirely clear, in particular in view of the different wording of the three equally binding languages of the EPC, it nevertheless can be stated with certainty that rodents or even mammals constitute a taxonomic classification much higher than species ('*Tierart*').[92] An 'animal variety' or '*race animale*'[93] is a subunit of a species and therefore of even lower ranking than a species. Accordingly the subject-matter of the claims to animals *per se* is considered not to be covered by the above three terms of the Article 53(b).[94]

Since, therefore, the altered mouse was not unpatentable by virtue of Article 53(b) first half-sentence, the case for considering whether or not it was a product of a microbiological process did not arise. The patent was granted. The case illustrates the difficulties patent-granting authorities face in interpreting complex legislative provisions. The Examining Division and Board of Appeal, both part of the EPO, reached different decisions on the same issues, supporting the view that Article 53(b) is difficult to interpret.

'*Essentially Biological*' Inventions

The second hurdle which Article 53(b) EPC presents for breeders is that plants or animals produced by means of an 'essentially biological' process are not patentable. What is meant by the term 'essentially biological'?[95] It has been suggested elsewhere three approaches to interpretation are possible:[96]

---

92   The German translation.

93   The French translation.

94   See Text of the Decision at point 2, [1991] EPOR.

95   The Patents Act 1977 UK amended by the Patents Regulations 2000 Schedule A2 defines 'essentially biological process' as 'a process for the production of animals and plants which consists entirely of natural phenomena such as crossing and selection'.

96   See Rainer Moufang, 'Patentability of Genetic Inventions in Animals', *supra* at 837.

1. All genetics belong to the field of biology and consequently all genetic processes are excluded from patentability.
2. Genetic inventions are not biological depending on how much the process involved relies on either physical or chemical means.
3. Legislators intended the term 'biological' to refer to 'natural' or uncontrollable processes. On this basis, the term 'essentially biological' extends to processes essentially without human intervention. However, if this were so the principle of 'disclosure' has no application in patent law in the sense that there would be no requirement of enablement in respect of repeatability for the skilled man.

If the third interpretation is the preferred option, the exclusion provision is without a separate, independent meaning.[97] Since the first approach to interpretation is unrealistic, the second approach is the preferred option. This accords with the approach adopted by the German Supreme Court in the 'Red Dove' decision.[98] In that case the court justified patenting of living-matter inventions by drawing an analogy between laws governing biological phenomena with those governing natural events for inanimate matter.

The Technical Board of Appeal considered the meaning of 'essentially biological' in *Lubrizol*. The Board said:[99]

> Fundamental alteration of the character of a known process for the production of plants is required before the applicant can avoid the rigors of being termed 'essentially biological'. The claimed processes must represent an essential modification of known biological and classical breeders processes.

Rainer Moufang argues[100] that this interpretation is an attempt by the Board to narrow the exclusion under the 'essentially biological' head, by regarding the provision as merely concretizing one of the patentability criteria, namely, inventive step. Support for Moufang's proposition is found in the judgment of the Board where it said:[101]

> Like any exception to a general rule of this kind the exclusion of 'essentially biological' processes for the production of plants (or animals) has to be narrowly construed. This is underscored by the fact that this exclusion does not apply to microbiological processes or the products thereof.

---

97   In the sense that it would merely concretize the repeatability requirement.

98   *Ex parte Schreiner* is reported at 1970 vol. 1 *International Review of Industrial Property and Copyright Law* 136–42.

99   See Reasons for the Decision at point 9.

100   See Rainer Moufang, 'Patentability of Genetic Inventions in Animals', *supra* at 838.

101   See Reasons for the Decision at point 6.

Later in the same judgment, the Board commented that an essentially biological process should be determined on the basis of the 'totality of human intervention and its impact on the results achieved'.[102] Implicit in the Board's comment is that an inventive step overrides the exclusion.

In *PGS*, the issue of whether or not claimed subject-matter was 'essentially biological' so as to remain unpatentable by virtue of Article 53(b) was also considered. There, the Board examined historical documents related to the drafting of Article 53(b) EPC, showing that patents should be granted for processes which, while applicable to plants, were of a technical nature. In *PGS*, the Board concluded that a process for the production of plants comprising at least one essential technical step, which could not have been carried out without human intervention, and where such intervention had such a decisive effect on the final result, was not excluded from patentability.[103] The Board, taking existing EPO jurisprudence into consideration, decided that the invention was not a result of a process which was 'essentially biological' in character.

Consensus within the EPO on a broad meaning of the term 'essentially biological' is welcome.

The 'Saving Clause' within Article 53(b)

Under Article 53(b) EPC Contracting States may not exclude micro-organisms from patent protection. The reason is because of the key role they play in the development of, in particular, new drugs. However, there is a delicate balance to be struck between meeting the needs of industry and broader social concerns such as avoiding an over-protectionist approach to the scope of patentable material.[104]

Article 53(b) second half-sentence permits the patenting of microbiological processes and the products thereof.[105] This represents an exception to the exclusion as set out in the provision. The lack of an express definition in patent law does not mean that microbiological is an undefined legal concept.[106] A sharp contrast can be drawn between inclusive material (where definitions are under-utilized) and excluded material (where definitions are used to limit the scope of protection). The emphasis on inclusion might be acceptable to those wishing to extend the limits of patentability but the apparent absence of controls over the extent to which this emphasis prevails adds to the concerns of those who favour a more restrictive

---

102   Ibid.

103   Ibid. at point 28.

104   See Mike Adcock and Margaret Llewelyn, 'TRIPS and the Patentability of Micro-organisms', 2000/2001 vol. 4 no. 3 *Bio-Science Law Review* 91–101 at 92.

105   The Patents Act 1977 UK amended by the Patents Regulations 2000 Schedule A 2 defines 'microbiological process' as 'any process involving or performed upon or resulting in microbiological material'.

106   See Mike Adcock and Margaret Llewelyn, 'TRIPS and the Patentability of Micro-organisms', *supra* at 96.

approach. How, therefore, is the clause to be interpreted? The Technical Board of Appeal in the *Onco-Mouse*[107] case was first to consider the scope of the clause. There, the Board said that the general principle of patentability under Article 52(1) EPC is restored for inventions involving microbiological processes and the products of such processes.[108] Consequently, patents can be granted over animals produced by a microbiological process. Therefore, it followed that animal varieties are patentable once they are the product of a microbiological process within the meaning of Article 53(b) EPC second half-sentence.[109]

The extent to which the saving clause benefits a breeder was again considered in the *PGS* case. There, the Board said that the exception provided for in Article 53(b) second half-sentence, which permits the patenting of microbiological processes and the products thereof, made it absolutely clear the EPC must provide patent protection for industrially applicable processes involving micro-organisms and products thereof.[110] The Board in *PGS* found support in the *Onco-Mouse* case, in which it was held that 'animal varieties' were patentable once they were products of a microbiological process. In *PGS* the Board considered that this principle applied *mutatis mutandis* to plant varieties.

According to the Board in *PGS*, although the EPC did not define 'microbiological processes or the products thereof', the provision was to be interpreted in accordance with objective teleological criteria, meaning that this way of interpreting is consistent with the legislative intent underlying the provision.[111] In this respect, the principle of 'equal treatment' is paramount. However, the Board in *PGS* did not give effect to its own ruling.

EPO jurisprudence at the time was to treat all generally unicellular organisms with dimensions beneath the limit of vision, and which could be propagated in a laboratory, as micro-organisms. The Board accepted this as defining 'micro-organism'. The Board said the term 'microbiological' was to be seen as qualifying technical activities in which direct use was made of micro-organisms. In other words, a technical activity in relation to a process or product[112] was microbiological only if such process or product involved micro-organisms as defined. Accordingly, the Board concluded that:[113]

> [t]he concept of 'microbiological processes' … [r]efers to processes in which micro-organisms as defined above, or their parts, are used to make or modify products,

107   [1991] EPOR 525.

108   Decision of Technical Board of Appeal, Oct 1990, T 19/90 3.3.2. See the Reasons for the Decision at 4.10.

109   See *PGS* [1995] EPOR 357, Reasons for the Decision at point 30.

110   This part of the provision was inserted to reflect states' practice of granting patents for industrial processes involving micro-organisms.

111   See Reasons for the Decision at point 32.

112   Meaning a process or product of manufacture.

113   See Reasons for the Decision at point 36

> or in which new micro-organisms are developed for specific uses. Consequently, the concept of 'the products thereof' encompasses products which are made or modified by micro-organisms, as well as new micro-organisms as such.

Nevertheless, the Board proceeded to hold that microbiological processes as defined and technical processes at least one essential step of which was of a microbiological nature could not be equated so that the principle of equal treatment applied.[114] The Board justified its decision by saying that the exception to the exclusion outlined in Article 53(b) second half-sentence referred to 'microbiological processes' and not to '*essentially* microbiological processes'. In *PGS* the subject-matter of the claim was unpatentable.

The decision of the Board is difficult to justify and offers little guidance for breeders who rely on patent protection to exploit microbiological processes and the products thereof commercially. The decision in *PGS* conflicts with EPO jurisprudence, namely, that exceptions to patentability should be narrowly construed. Indeed, in this respect, the Board is in conflict with itself. It is difficult to foresee a situation in which the exception to the exclusion outlined in Article 53(b) can be applied, if not in the *PGS* case. Therefore, notwithstanding the decision of the EBA in *Novartis*, it seems that the practice of providing patent protection over a broader rather than narrower range of biological material is far from settled.

Article 53(c)

As mentioned earlier, as a result of the revised EPC 2000, under Article 53(c) methods of treatment or diagnosis of the human or animal body are now precluded from patentability in their own right. This is for public policy reasons. Such inventions are no longer precluded on the basis of the fiction that they are not industrially applicable. While no change of practice is expected to arise from the amendment, the policy shift is significant and shows that the exceptions to patentability outlined in Article 53 are merely illustrative and not exhaustive. By contrast, in Australia public policy issues are not determinative of patent applications. In *Anaesthetic Supplies Pty Ltd v Rescare Ltd*[115] it was held the law permitted the grant of a patent for a method of treatment of human beings for the alleviation of disease, malfunction or incapacity because there was no logical distinction in principle between patenting a product for treating the human body and a method of treating the human body. In that case it was said: 'Australian courts must now take a realistic view of the question in light of the current scientific and

---

114   Ibid. at point 38.
115   *Anaesthetic Supplies Pty Ltd v Rescare Ltd* [1993/94] Federal Court of Australia.

legal developments … and the courts should be hesitant to introduce an exception based on very general principles of ethics and social policy.'[116]

The conclusion is that under the revised EPC 2000, exceptions to patentability are no longer to be determined solely on (clear) patentability criteria but also on principles of ethics and social policy, thus adding to the confusion already experienced by biotechnologists in patent applications.

Conclusions

EPO jurisprudence is unclear in respect of what is unpatentable by virtue of Article 53(a) EPC. The *Onco-Mouse*[117] case established the 'balance of interests' test whereby patentability is not precluded on the basis of morality where the 'benefit' outweighs the 'cost' of the invention. The *Howard Florey/Relaxin*[118] case established that patentability is precluded only where the general public considers the invention so abhorrent as to be inconceivable. The *PGS*[119] case established that the test laid down in *Onco-Mouse* is not appropriate in all circumstances. What the cases demonstrate is that Article 53(a) may no longer protect adequately biotechnological inventions. Patentees are entitled to know when their actions are safe and lawful. Yet too much uncertainly exists. This supports the view Article 53(a) is not an appropriate mechanism for determining exceptions to patentability on moral grounds.

What EPO jurisprudence does make clear is that morality is now a matter for European institutions. The significance of this has yet to be worked out in the context of national law. However, until then, it is likely that EPO jurisdiction will continue to expand beyond that envisaged in the EPC by virtue of the creation of Article 53(a). In the context of morality the application of Article 53(a) EPC is likely to result in inconsistency of practice among EPC states. Unless the Office offers clear evaluation criteria in respect of *ordre public* and morality, it is unable to offer sufficient guidance for future cases, suggesting that lack of uniformity among states is likely to continue.

Likewise, EPO jurisprudence is unclear in respect of what is unpatentable by virtue of Article 53(b) EPC. Insistence by the EPO upon strict construction in relation to Article 53(b) means that:

- Plant varieties receive no protection at all in certain circumstances.
- There is no legal regime of animal protection.
- A dual system of protection exists in respect of living matter inventions.

---

116   Per Lockhart and Wilcox JJ.
117   [1991] EPOR 525.
118   [1995] EPOR 541.
119   [1995] EPOR 357.

EPO jurisprudence is unclear in respect of subject-matter patentability and the application of Article 53(b) EPC is likely to result in inconsistency of practice among EPC States. The cases demonstrate it is not an appropriate mechanism for determining exclusions from patentability. What is required is a policy clearly outlining the scope of protection and the grounds and extent of exclusion of certain subject-matter from patentability.

The policies underlying the creation of Article 53 EPC are, or have become, unclear with the result that biotechnological inventions no longer receive adequate protection. What is required is effective harmonizing legislation as regards patentability criteria, including a consensus on the part of states as to the place of morality within the patent system.

Chapter 5
Patent Law Criteria and Biotechnological Inventions

Introduction

Biotechnological inventions, in particular genetic engineering, involves the discovery of new biological processes or the alteration of existing ones. The basis of many biotechnological inventions is informational. The description of a biotechnological invention, therefore, focuses on a modification of existing complexity. Man's inability to create living matter per se shows the limited understanding man has of living processes. If the day arrives when man can fully understand the complexity of living matter and how it is created, then a description of biotechnological inventions may be grounded in terms of the organizational principles of living matter. However, until that day arrives, descriptions of biotechnological inventions will be based on approximations of such organizational principles.[1] That is to say, biotechnological inventions are likely to be defined more appropriately in functional or informational terms rather than structure. In this sense in particular, biotechnological inventions have proved challenging for the disclosure requirement of patent law.

To protect adequately biotechnological inventions what is required is effective harmonizing legislation as regards patentability criteria. In this regard, for European states international agreement on substantive patentability criteria outlined at Strasbourg and later incorporated into the European Patent Convention (henceforth EPC) – and the Directive – may not in itself be sufficient to result in harmonization. In addition, a consensus on the part of states as to the place of morality within the patent system is also required. However, some patent-granting authorities – in particular, it would seem, in the UK – interpret traditional criteria in a manner, which, is not conducive to protecting biotechnological inventions.

This chapter outlines how, under the EPC and the Patents Act 1977, the cases demonstrate that sufficient difficulties already arise when traditional substantive patent law criteria, namely, disclosure, novelty, inventive step and industrial application, are applied to fast-moving, emerging technologies such as biotechnology without morality being presented as a general criterion for patentability. Because these criteria are not conducive to patenting biotechnological

---

1 See generally Stephen A Bent, Richard L Schwaab, David G Conlin and Donald D Jeffery, *Intellectual Property Rights in Biotechnology Worldwide* (1987) Stockton Press, NY and Macmillan Publishers, London.

inventions, in particular in the UK, this suggests that other problems created in addition by moral constraints on the margins are undesirable. Such problems include: establishing what the aims of a moral policy are; how such aims are achieved by the incorporation of a moral policy into patent law; and whether, and to what extent, moral policy compromises economic policy. Although the policies underlying the traditional criteria are firmly established, the reliability of tests is uncertain nevertheless. This suggests that, because the aims of a moral policy are unclear, the acceptability of tests and the ability of examiners to rule on matters moral are questionable.

Neither the European Patent Convention 1973 nor the Patents Act 1977 is drafted with the special characteristics of biotechnological research in mind.[2] Therefore, if such research is to receive the protection it deserves, the concepts underlying traditional patent law may need to be adapted. The cases, discussed hereafter, suggest that the European Patent Office (henceforth EPO) is more willing than UK courts to do this. The EPO granted a patent to both Genentech[3] and Biogen[4] for certain biotechnological inventions. By contrast, neither company was awarded a patent for the same inventions in the UK. Thus, a comparative perspective is offered. It is the author's contention that, because traditional substantive patent law provisions are already problematic enough for biotechnological inventions, in terms of protection, moral considerations on the margins of the system are unhelpful.

In this section UK Patent Office jurisprudence is contrasted with that of the EPO to determine whether or not the EPO is more willing than UK courts to adapt traditional patent law criteria to protect biotechnological inventions, and, if yes, on what basis. The problems presented by the language and concepts of the Patents Act 1977 for courts in the UK are outlined, as are those created for the EPO by virtue of the EPC. Difficulties which substantive patent law criteria pose for biotechnological inventions, in particular the concepts of obviousness and inventive step, who is the man skilled in the art and the disclosure requirement, are examined.

Establishing a common system for the grant of patents required harmonization of certain legislative provisions and, as discussed in Chapter 2 heretofore, resulted, first, in the Strasbourg Convention 1963 and, later, in the European Patent Convention 1973. Effect was given to the EPC in the UK by means of the Patents Act 1977, section 130(7) of which tells us that certain parts of the Act are framed 'so as to have as nearly as practicable the same effects as their counterparts in the Convention'.[5] Mentioned specifically in this connection are,

---

2   See, however, the Patents Regulations 2000, which amend the Patents Act 1977 in order to make provision for the Directive on the Legal Protection of Biotechnological Inventions 1998.

3   *Genentech 1/Polypeptide Expression* [1989] 1 EPOR 1.

4   *Biogen/Hepatitis B* [1995] EPOR 1.

5   In light of the Patents Regulations 2000, the amended Patents Act 1977 is likely to have the same effects as counterparts in the Biotechnology Directive 1998 – thus,

inter alia, sections 5(2), 14(3), 14(5) and 72(1), which deal with issues of priority, disclosure and revocation respectively. The thread running through these sections is that of 'enabling disclosure'.

For the inventor the 'price' of obtaining a temporary monopoly is that the invention must be disclosed sufficiently to enable the skilled man in the relevant art to carry out the invention. The principle of sufficient disclosure is premised on public policy considerations. Hence its importance. The rationale behind the principle of disclosure is that, while the inventor should be rewarded for his effort, competition in the relevant art must not be eliminated.

As the cases testify, courts in the UK have struggled with issues of policy in respect of traditional patentability criteria concerning biotechnological inventions.[6] Whether or not patent law should extend to biotechnological inventions at all seems problematic. However, denying monopoly protection to those who have invested heavily in terms of time, expertise and money can discourage workers in the field from making advances of the greatest public benefit. The concern of competition is to ensure that, in the interests of the public, research is not stifled. The reasoning of the conflict between granting a monopoly and preserving competition may be inadequate in genetic engineering cases because to grant the patentee a monopoly in the product of his work is not to give effective protection from competition – since, once other routes to that product become known, competitors can manufacture without incurring a large portion of the initial cost of the project.

In contrast to the protection afforded by the EPO, in the author's opinion, the UK cases demonstrate that biotechnological inventions do not always receive the protection they deserve. There are two reasons why such a situation exists. First, decisions of the courts tend to be dominated by policy issues that are not accommodated within the legislative framework of the Patents Act 1977. This can result in biotechnological inventions being excluded from the scope of patent protection. Second, the language[7] and concepts of the Act as understood against the background of British patent decisions have influenced the judiciary such that they are reluctant to cope with new problems associated with biotechnological inventions.[8]

---

harmonizing certain legislative provisions.

6   See Iain Purvis, 'Patents and Genetic Engineering: Does a New Problem Need A New Solution?', 1987 vol. 12 *European Intellectual Property Review* 347–8.

7   In no area of law is there a greater temptation to use science and scientific terminology than in patent law when defining an invention and interpreting the meaning of language and the boundaries of that language in ever-changing circumstances. Modern biotechnological inventions and (modern) patent law hide behind scientific lines of demarcation in an effort to resolve difficult legal questions. See Robin Feldman, 'The Role of Science in Law', Ch. 1 in *The Internationalisation of Science in Modern Law* (2009) at 28, Oxford University Press.

8   The language of patent law is not suited for living organisms because it requires their animate nature be reduced to chemical descriptions which are misleading because they suggest we have a complete materialist explanation of life, and we do not. See Kathryn

Disclosure and the Patent System

Disclosure lies at the heart of the patent system and, in this respect, is treated as a substantive criterion of patent law. The problems of the policies underlying the disclosure provision, and whether or not they are accommodated by the EPO and UK Patent Office to protect adequately biotechnological inventions, are examined hereafter. The relationship between the issue of morality and issues relating to the disclosure of the invention is one of contrast. Despite being problematic, the aims of the policy underlying the disclosure requirement are clear; this is the reason why its place within the patent system is firmly established. By contrast, the policies underlying the creation of Article 53 EPC are, or have become, unclear and, in the author's opinion, no longer protect adequately such inventions. The application of the disclosure requirement and whether or not it is adapted to protect biotechnological inventions, in a manner that the morality provision is not, is outlined hereafter.

By virtue of the EPC an invention is firstly and mainly disclosed in the description of the invention's patent application as required by Article 83 thereof. It is additionally disclosed in the claims and in the drawings of the patent application as required by Article 84 EPC.

Article 83 EPC formulates the sufficiency of disclosure requirement as follows:

> The European patent application must disclose the invention in a manner sufficiently clear and complete for it to be carried out by a person skilled in the art.

The equivalent provision in the Patents Act 1977 UK is section 14(3). It reads as follows:

> The specification of an application shall disclose the invention in a manner which is clear enough and complete enough for the invention to be performed by a person skilled in the art.

The support requirements are outlined in Article 84 EPC and read as follows:

> The claims shall define the matter for which protection is sought. They shall be clear and concise and be supported by the description.

The equivalent provision in the Patents Act 1977 is outlined in section 14(5). It reads partly as follows:

Garforth, 'Life as Chemistry or Life as Biology? An Ethic of Patents on Genetically Modified Organisms', in Johanna Gibson (ed.), *Patenting Lives: Life Patents, Culture and Development* (2008), at 52, Ashgate Publishing, UK

The claim or claims shall:

(c) be supported by the description.

The description requirements are outlined in the Implementing Regulations EPC, according to Rule 27(1)(a–f), a patent application must:

(a) specify the technical field

(b) indicate the background art

(c) disclose the invention in such terms that the problem and solution can be understood, and state any advantageous effects

(d) briefly describe the figures in the drawings

(e) describe in detail at least one way to carry out the invention using examples

(f) indicate explicitly if necessary, the way in which the invention is capable of exploitation in industry.

Under traditional patent law doctrine, the disclosure requirement serves three functions.[9] First, it ensures identification of the invention. Second, it ensures that the invention is reproducible. Third, it ensures that the invention is industrially applicable. Such is the importance of the disclosure requirement that non-conformity serves as a ground for revocation. Article 138(1) EPC reads partly as follows:

Subject to the provisions of Article 139, a European patent may only be revoked under the law of a Contracting State, with effect for its territory, on the following grounds:

(b) if the European patent does not disclose the invention in a manner sufficiently clear and complete for it to be carried out by a person skilled in the art.

The equivalent provision in the Patents Act 1977 is section 72(1). It reads partly as follows:

Subject to the following provisions of this Act, the court or the comptroller may on the application of any person by order revoke a patent for an invention on (but only on) any of the following grounds, that is to say

---

9   See generally E S Van de Graaf, *Patent Law and Modern Biotechnology* (1997) 309–52 Gouda Quint, The Netherlands.

(c) the specification of the patent does not disclose the invention clearly enough
and completely enough for it to be performed by a person skilled in the art.

As already noted, certain sections of the Patents Act 1977 UK are framed to
follow closely parts of the EPC. This suggests that, in terms of protection, these
sections should have the same effect as their counterparts in the EPC. Therefore,
it is reasonable for the patentee to rely on uniform and consistent jurisprudence
between the UK and the EPO. However, the cases do not support this.

Disclosure and European Patent Office Jurisprudence

EPO jurisprudence suggests that policy issues relating to disclosure have largely
been accommodated within the existing legislative framework of the EPC. The
reason is that the aims of the policies are clear. In its decisions the EPO has
demonstrated a willingness to interpret the language and concepts of the EPC in a
manner affording patent protection to biotechnological inventions.

Genentech 1/Polypeptide Expression

In the case of *Genentech 1/Polypeptide Expression* (henceforth *Genentech 1)*[10] the
Technical Board of Appeal overturned the decision of the Examining Division and
allowed the appeal. Detailed discussion of the case is justified on the basis that, in
pioneering technology, unless a broad view of invention is taken, entrants into the
field will not be encouraged.

   In EPO jurisprudence, this case has led the field on the subject of Article 83
and to this extent it is innovative. In the case the EPO examined the relationship
between Article 83 and Article 84 EPC. Conflicting decisions by EPO Divisions
show the difficulties involved when patent-granting authorities apply traditional
Convention provisions to biotechnological inventions.

   The patent in suit concerned a human growth hormone. Essentially, the claim
related to a recombinant plasmid (a self-reproducing particle of protoplasm,
without a nucleus, which can determine hereditary characteristics) into which
artificially made DNA was inserted. The DNA encoded a desired functional
polypeptide. When grown in a suitable bacterium, the desired polypeptide was
expressed and recovered. Claim 1 was drafted in broad terms and read partly as
follows: 'A recombinant plasmid suited for transformation of a bacterial host
wherein the plasmid comprises a homologous regulon, heterologous DNA, and one
or more termination codons.' The invention was said to reside in the ability of the
restructured plasmid, when inserted into suitable bacteria, to *control* the expression
of the polypeptide. Genentech claimed the expression of the polypeptide using
any modified plasmid. The modified plasmid comprised a regulon, an expression

---

10   [1989] 1 EPOR 1.

control sequence composed of amino acids. Genentech claimed other expression control sequences whose functions were as yet unknown. In short, Genentech claimed variants to produce the required polypeptide.

The Examining Division rejected the application. The stated grounds for refusal included insufficient disclosure to meet the requirements of Article 83 EPC, and lack of support to meet the requirements of Article 84 EPC. The Division insisted that all embodiments of the invention as set out in the claim (in conformity with Article 84) were capable of performance by the skilled man in a repeatable manner without practising inventive skill. Since some of the claimed sequences remained unknown, certain embodiments of the invention as claimed could not be reproduced without some form of further invention involving identification of the sequences. For the Division, therefore, the application fell under Article 83 because the invention was not disclosed sufficiently to enable the skilled man to carry it out without practising further inventive skill. The Division proceeded to hold that the application also fell under Article 84 for the following reason: the claims described the invention by what it did. According to the Division an invention defined by what it did, rather than what it was, could not 'define the matter' for which protection was sought.

The Technical Board of Appeal disagreed with the decision of the Examining Division in respect of Article 83 and Article 84 and allowed the appeal.

In respect of Article 83, the Board held that, in cases representing biological processes, the test as far as sufficiency was concerned was whether or not the process as such was *reproducible*. The decision of the Board suggests that once the process was reproducible the skilled man necessarily possessed sufficient instruction by means of the disclosure to work the invention. The test of reproducibility laid down by the Board was eminently suitable for biotechnological inventions because it highlighted the codependence of the disclosure and support requirements in the following manner. Once reproducibility is established, identification of the invention is guaranteed. This means that the identifying disclosure required under Article 83 affects the scope of the invention required under Article 84. That is to say, the purpose of Article 83 is to ensure that the invention is sufficiently disclosed, and that of Article 84 is to ensure the claims define the invention for which protection is sought. Since the test of reproducibility guarantees identification of the invention, claims under Article 84 are necessarily limited to those related to that invention. Because a relationship subsists between Article 83 and Article 84, the nature of the claim is unimportant. The test is not as to the type of claim, but rather its extent. Functional type claims are acceptable simply because, like other types of claim, what matters is the extent of the monopoly embraced within the claim. The issue is: how can the subject-matter of the claim be disclosed so as to satisfy the requirements of Article 83? The reproducibility test adopted by the Board set the parameters which defined, not only the extent of the monopoly claimed, but also the technical contribution to the art. The legal basis of the test was rooted in Article 83 and Article 84. Worked together, the two articles sought to ensure, not only that research was encouraged, but that competition was preserved. Although not expressly stated, *implicit* in the judgment is that disclosure for Article 83 purposes must be sufficient to enable performance across the full width of the invention as claimed. In this manner the extent of the monopoly can be defined.[11] The rejection of claims under Article 83 is grounded in relevant features for the identification of the *claimed invention as a whole*. In other words, *essential* claim features are relevant for Article 83 purposes. Reproducibility requires also that the invention be obtained regularly. In this manner the contribution to the art is defined. This suggests that disclosure related to both availability of starting materials and repeatability of consecutive steps of the claimed invention.

In respect of Article 84, the Board cited with approval the case of *Synergistic Herbicides*,[12] in which functional terminology was approved. In this case, it was held that where the features of an invention could not otherwise be defined more precisely *without restricting the scope of the invention*, then functional terminology could be employed. Hence, the Examining Division's argument, that functional claims did not define the matter for which protection was sought, was rejected by the Board.

Defining the scope of the invention was, therefore, of vital importance. The Board took the view that the invention was akin to a principle, namely, the ability to express DNA in polypeptide form using suitably engineered plasmids and suitable bacterial hosts. In order not to restrict the scope of the invention, it was necessary for Genentech to claim functional features. The claims could therefore embrace features related to plasmids, regulons and bacteria. Functional variants (unknown or which might be provided for in the future by way of further invention) should not be excluded from the scope of the claims, as long as the claims contained some variants which were known to the skilled man, either through the disclosure or common general knowledge.[13]

The approach adopted by the Board to the issue of inventive scope was appropriate for biotechnological inventions. The Board saw the invention partly in the ability to transform suitable bacteria for the purposes of polypeptide expression. Until transformed, such bacteria remained unknown. By analogy, therefore, the Board was willing to see the invention partly in the ability to transform plasmids by the insertion of regulons, notwithstanding that the amino acid sequences of such regulons, as yet, remained unknown.

The Board distinguished between essential and non-essential claim features of an invention. Essential claim features were those required to be known in order to enable the skilled man to reproduce the invention. To afford proper protection to the scope of the invention, therefore, it was necessary that variants of essential components also received protection. The Board said:[14]

---

11   Whether or not this case supports so wide a proposition is debatable. See R Stephen Crespi, 'Biotechnology, Broad Claims and the EPC', *supra*.

12   Case no. T68/85, O J 6/1987 228 at para. 3.3.1.

13   See *Genentech 1* [1989] *supra* at para. 3.1.5.

14   Ibid.

Unless variants of components are also embraced in the claims, which are, now or later on, equally suitable to achieve the same effect in a manner which could not have been envisaged without the invention, the protection provided by the invention would be ineffectual. Thus it is the view of the Board that an invention is sufficiently disclosed if at least one way is clearly indicated enabling the skilled person to carry out the invention.

The test of reproducibility laid down by the Board suggests that all features required for defining the invention for Article 84 purposes must be disclosed. What Article 83 required was that the essential claim elements were sufficiently disclosed to enable the skilled man to work the invention. The test of reproducibility, incorporating as it does identification of the invention, ensured sufficient disclosure. Disclosure of how to work the invention using all possible claimed variants was not a requirement of Article 83. For the Board, disclosure and support were elements governed by public policy considerations. The Board said: '[t]he need for a fair protection governs both the considerations of the scope of the claims and of the requirements of sufficient disclosure'.[15]

The case demonstrates that in a young science such as biotechnology, where new accomplishments are achieved for the first time, the technical contribution deserves to be better recognized, or recognized more fully. In reasons for its decision, the Board showed that it was willing to interpret the disclosure requirements of the EPC in a flexible manner. The decision is proactive in respect of biotechnological inventions and shows three things. First, policy issues can be implemented without excluding biotechnological inventions from the scope of patent protection. Second, the patent system is designed to afford protection to all types of invention. It can just as easily cope with biotechnological as with mechanical inventions. Third, in respect of patentability criteria the Technical Board is willing to adopt a flexible attitude.

Biogen/Hepatitis B

In *Biogen/Hepatitis B* (henceforth *Biogen*),[16] the EPO again adopted a flexible approach to interpreting traditional patent law criteria in order to protect new technology. Here, the Board considered the disclosure requirements of the EPC and how they are satisfied in relation to inventions claiming priority from earlier filed applications. The case is of particular importance because the decision of the EPO, which upheld the patent, conflicts with the UK courts' revocation of the patent (discussed hereafter).[17]

---

15   Ibid.

16   [1995] EPOR 1.

17   See the decision of the House of Lords in *Biogen Inc. v Medeva PLC* [1997] RPC 1, per Lord Hoffmann, confirming the decision of the Court of Appeal in revoking the patent.

Legal background   The case deals with technically complex issues. Briefly, the principal claims of the patent related to an artificially constructed molecule of DNA carrying a genetic code which, when introduced into a suitable host cell, caused that cell to make antigens of the virus hepatitis B.[18] Hepatitis B antigens are of two types: core antigen and surface antigen. Hepatitis B is a widespread human virus whose antigens can be used both to test whether someone has the virus, and to make a vaccine giving immunity against infection.[19] Biogen claimed to be able to express hepatitis B viral antigens of both the core and surface type, using any method of recombinant DNA technology.

The patent in suit was granted on an application filed in the EPO on 21 December 1979[20] (henceforth EP 793) claiming priority[21] (under Article 87(1) EPC) from an application (referred to as Biogen 1) filed in the UK on 22 December 1978. The 'priority date' of a patent is the date on which it is tested against the 'state of the art'. It is the date on which it, or any application claiming priority from it, becomes part of the art, when assessing the novelty of later applications. In order to claim priority, the invention to which the application in suit relates must be supported by matter disclosed in the earlier application. The EPO published the application on 11 July 1986, and granted the patent on 11 July 1990. Opposition was entered in April 1991. The Opposition Division held that the patent was invalid. On appeal, the Technical Board upheld the validity of the patent.

Technical background[22]   In 1970 a paper was published by D S Dane,[23] which suggested that the infective agent for hepatitis B virus was a certain particle consisting of a circular molecule of DNA in a protein core, which was surrounded by surface proteins. This later became known as the Dane particle. The infection was accompanied by an overproduction of particles made up only of the surface protein. The surface proteins did not contain hepatitis B viral DNA and were therefore non-infective. Medical preparations of surface antigen taken from patients could be used as vaccines (so-called 'first generation' vaccines) but this form of preparation was limiting. Consequently, this method of treatment did not find favour with the medical profession.

---

18   Antigen is a substance which can provoke the formation of an antibody when introduced into another organism, e.g., a vaccine can act as an antibody.

19   See *Biogen Inc. v Medeva PLC* [1997] *supra* per Lord Hoffmann at 32, where he outlined some of the technical background.

20   European Patent Application no. 79303017.2.

21   See generally W R Cornish, *Intellectual Property: Patents, Copyright, Trade Marks and Allied Rights* 3rd edn (1996) Sweet & Maxwell, UK.

22   See *Biogen Inc. v Medeva PLC* [1997] *supra* per the judgment of Lord Hoffmann at 34 *et seq.*

23   Dane et al., *Lancet* 1970 at 695–8.

A possible alternative to this method of treatment was to make antigens artificially by chemical synthesis. In 1977 Paterson[24] identified a sequence of nine amino acids at the end of the chain of the surface protein. However, his discovery did not result in the synthesis of antigen.

Another alternative was the use of DNA technology. Two possible approaches existed. The first was to identify that part of the viral DNA coding for the surface antigen and to express that. This would have required the complete sequencing of the viral genome and would take a long time. The other was to express random pieces of the DNA to see what was produced. This involved splicing the viral DNA at random, along with cloning and expressing the random samples, to see whether DNA coding for any useful proteins had been selected. This was a high-risk strategy with a good chance of failure. Nevertheless, it was the approach Professor Murray[25] adopted in 1978 when researching for Biogen. Biogen sought to express antigen in a suitable host cell, namely, that of the bacterium *E. coli*. One of the many difficulties involved, therefore, was the expression of *unsequenced* eukaryotic DNA in a prokaryotic host (i.e., the expression of human DNA in a bacterial host). It was this particular difficulty which made the invention 'not obvious to try'. It was generally believed to be true that, like other viruses, *E. coli* contained 'introns' (intervening nucleotide sequences on the DNA not forming part of the primary code for the polypeptide and not expressing genes),[26] consequently the skilled man in the art would not have thought expression worth trying. It was Biogen's contention that, because their decision involved trying what was in effect not obvious, it was sufficient to amount to an inventive step.

Claim 1 of the patent in suit read partly as follows: 'A *recombinant* DNA molecule *characterised* by a DNA sequence coding for a polypeptide or a fragment thereof displaying HBV [hepatitis B virus] antigen specificity.' Claim 1 was a product claim involving a molecule identified partly by the way it was made (recombinant DNA technology), and partly by what it did. The claim generalized what had been done in two ways: first, as to the results to be achieved, which covered molecules capable of producing both core and surface antigen; and, second, as to the method used in achieving the results, which covered both bacterial and non-bacterial hosts.

The case before the Technical Board centred on entitlement to priority. The issue was: could EP 793 claim priority from Biogen 1, the earlier British-filed application? In order to do so, EP 793 and Biogen 1 needed to relate to the same invention. In effect, the invention claimed in EP 793 ought to have been disclosed in the earlier Biogen 1 document in such a manner as to support the later invention.

24   Paterson et al., 1977 vol. 74(4) *Proceedings of the National Academy of Science of the United States of America* 1530–34.

25   Professor Sir Kenneth Murray of Edinburgh University.

26   See James M Marshall, 'Biotechnology Patents: A Further Twist', 1998 Feb *Patent World* 25–8 at 28.

The Technical Board looked to Article 87 EPC, which is concerned with entitlement to priority. It reads as follows:

> A person who has duly filed in, or for, any State party to the Paris Convention for the Protection of Industrial Property, an application for a patent ... [s]hall enjoy, for the purposes of filing a European patent application in respect of the same invention, a right of priority during a period of twelve months from the date of filing the first application.

Therefore, there is a period of 12 months from which the later application can claim priority from the first application, as long as both relate to the same invention; the later invention must be disclosed in the prior document sufficiently clearly to support the later application.

In the view of the Board,[27] the main criterion in this respect was whether or not the claimed invention was disclosed in the priority document as a 'matter of substance'. According to the Board, disclosure of the essential elements of the invention 'must be either express, or directly and unambiguously implied by the text'.[28] The Board examined the subject-matter of the Biogen 1 priority document.[29] It was satisfied that the document referred to how fragments of viral DNA were inserted into *E. coli*, thereby allowing expression of polypeptides of the antibody variety. The Board said that the document related to the transformation of host micro-organisms, with vectors containing HBV DNA, appropriately cleaved and inserted into hosts, so that they produced polypeptides with the specificity of HBV antigens. The Board said: '[i]t is directly and unambiguously implied by the text that the expressed product could be a protein displaying the antigen specificity and antigenicity of one or more of the known HBV antigens'.[30] The prior Biogen 1 document, therefore, made an explicit reference to all the essential elements claimed in the later invention.

For the Board, the production of antigen was crucial. This was an essential element of the invention. The invention was embodied in the production of either core or surface antigen. The lack of data on the actual production of one or other form of antigen was not crucial, and did not necessarily lead to the conclusion that essential elements of the claimed invention were missing from the priority document. For the Board, the later invention was sufficiently disclosed in the priority document once the priority document supported the later invention by satisfying the test of reproducibility as a matter of substance. Reproducibility guarantees identification of the invention. Identification necessarily embraces all the essential features of the invention.

---

27   See *Biogen/Hepatitis B* [1995] EPOR 1 at para. 4.1 of the Reasons for the Decision.

28   Test laid down in *Collaborative/Preprorennin*, case T81/87 [1990] EPOR 361.

29   See *Biogen/Hepatitis B* [1995] *supra* at para. 4.3.

30   Ibid. at para. 4.5.

The decision of the Board can be subjected to criticism. The expression of one form of antigen, but not the other, cannot fairly be said to identify the invention. If one accepts the proposition of the Board, namely, that invention was identified by the expression of one form of antigen, then expression of either form of antigen must surely be regarded as an essential element of the invention. If the invention is embodied in the expression of either of two forms of antigen, then both forms must be disclosed. The test of reproducibility applied by the Board in the instant case failed to identify the invention as claimed. Notwithstanding such criticism, the decision of the Board was welcome. The Board showed that it was willing to take a broad view of invention, thus encouraging entrants into the field. The priority document claimed monopoly over expression of either core or surface antigen. This was justified by the contribution to the art. However, the view has been expressed by other judicial authorities (discussed hereafter) that the Board failed to address the question of whether or not the claims exceeded the contribution to the art.[31]

Whatever the merits or demerits of the actual decision, the judgment demonstrates that, notwithstanding the flexible approach adopted by the EPO, sufficient difficulties arise when traditional substantive patent law provisions are applied to biotechnological inventions such that other criteria on the margins are not helpful. Given that the policies underlying the issue of disclosure are clear, doubt must exist regarding the ability of patent examiners to rule on matters moral where policy considerations are less clear.

Disclosure and UK Jurisprudence

By contrast with the EPO, UK courts are not flexible enough when applying traditional patent law principles, particularly in regard to disclosure, to accommodate new technology. Because certain parts of the Patents Act 1977[32] are framed to have as nearly as practicable the same effects as their counterparts in the EPC, judicial authorities in the UK often look to EPO jurisprudence for guidance on how to interpret these provisions. This has resulted in confusion between lower and superior courts in the application of the disclosure requirement in the UK. The cases *Biogen Inc. v Medeva PLC*[33] (henceforth *Biogen*) and *Genentech's Inc.'s Patent*[34] (henceforth *Genentech*) demonstrate, in the context of disclosure, the approach of the higher courts in England towards biotechnological inventions. The cases[35] reveal that, in the context of patentability criteria, the language of the Patents Act 1977 is not conducive to protecting biotechnological research. As

---

31 See *Biogen Inc. v Medeva PLC* [1997] *supra* per the judgment of Lord Hoffmann.

32 Amended by the Patents Regulations 2000. In force 28 July 2000.

33 [1997] RPC 1.

34 [1989] RPC 147.

35 Not discussed in chronological order; this is to facilitate arguments as they arise.

already noted, both companies were awarded patents for the same invention by the EPO. These and related issues are examined in this section.

Biogen Inc. v Medeva PLC

The case of *Biogen Inc. v Medeva PLC*[36] was first heard in the Patents Court before Aldous J and concerned an action for infringement of a European Patent (UK) relating to hepatitis B virus in which the defendant counterclaimed for revocation of the patent. One of the issues Aldous J had to deal with was whether or not the invention to which the application in suit related was supported by matter disclosed in the earlier application of Biogen 1.

The test for priority is laid down in section 5(2)(a) of the Patents Act 1977. It reads as follows:

> If an invention to which the application in suit relates is supported by matter disclosed in the earlier relevant application or applications, the priority date of that invention shall instead of being the date of filing the application in suit be the date of filing the relevant application in which that matter was disclosed or, if it was disclosed in more than one relevant application, the earliest of them.

In the Patents Court Aldous J said that section 5(2) required the court to decide whether the invention claimed in claim 1 and in any other claims was supported by matter disclosed in Biogen 1. He referred to the *Asahi* case,[37] in which Lord Oliver had said that a description in an earlier application which did not contain *enabling disclosure* would not support the later invention, so as to enable it as an invention to claim priority from the date of that application.[38] The issue before the Patents Court, therefore, became whether Biogen 1 supported the invention in the application in suit sufficiently to satisfy section 5(2). This depended on how the term 'invention' was construed by the court. In brief, for Aldous J the invention was the production of antigen and his resolution on the issue of priority was relatively straightforward. Medeva admitted in the proceedings that the specification disclosed sufficient instructions to enable the skilled man to express core antigen. Therefore, the allegation of insufficiency by Medeva could not succeed, since Aldous J had already ruled that claim 1 related to one invention only. Aldous J concluded that there was compliance with the requirements of section 5(2). Biogen 1 did support the invention in the application in relation to the patent in suit.

The Court of Appeal disagreed with Aldous J and decided that claim 1 related to more than one invention. Therefore, if both were to have the benefit of priority, support must be found for both in the earlier application of Biogen 1. Hobhouse

---

36   [1997] RPC 1.
37   *Asahi Kasei Kogyo KK's Application* [1991] RPC 485.
38   Ibid. per the judgment of Lord Oliver at 536.

LJ, giving judgment for the court, said that support included the requirement of section 14(5)(c) of the Act, namely, the claim must be supported by the description. In regard to sufficiency of disclosure, it followed by virtue of section 14(3) of the Act that support must include disclosure of the inventions in a manner clear enough and complete enough for the inventions to be performed by someone skilled in the art. The view of the Court of Appeal was that Biogen 1 merely taught how to express core antigen in a bacterial host. The claim extended to both core and surface antigen in bacterial and non-bacterial hosts. Since, therefore, Biogen 1 did not teach how to make surface antigen, the later application relating to the patent in suit could not claim the benefit of priority from Biogen 1. Hobhouse LJ said: 'Support is a matter of substance not mere form. In so far as the matter disclosed by Biogen 1 does not suffice to support the invention or inventions, they are not entitled to the priority of Biogen 1.'

In effect, what the Court of Appeal said was that it did not matter whether the production of antigen was seen as either one or two inventions. Priority could not be claimed in either event. If the production of antigen was seen as a unitary invention involving the expression of core antigen, then Biogen 1 merely taught how to make this. However, the expression of surface antigen was also an essential element of the invention and, therefore, needed to be disclosed. Consequently, disclosure of 'one way' of performing the invention was not enough. If the production of antigen was seen as two inventions, then disclosure of both was necessary.

The Court of Appeal decision did away with the so-called 'one way' rule. Disclosure to be sufficient required the skilled man to carry out the invention *as claimed*. Therefore, where the claims related to more than one embodiment of the invention, disclosure of how to work such embodiments must be made.

The House of Lords upheld the decision of the Court of Appeal in relation to the 'one way' rule. The House referred to, and confirmed, the decision of Lord Oliver in *Asahi*. However, what was less clear according to the House was what exactly the concept of enabling disclosure meant. Lord Hoffmann considered this to be the critical issue in the case.[39] He said: 'It is not whether the claimed invention could deliver the goods, but whether the claims cover other ways in which they might be delivered: ways which owe nothing to the teaching of the patent or any principle which it discloses.' For Lord Hoffmann, the issue was not whether the method put forward by Professor Murray[40] could produce both core and surface antigen, but whether what the Professor sought to protect covered other ways in which the antigens might be produced; ways not taught by his work. Lord Hoffmann was prepared to give protection for the production of antigen using unsequenced DNA but not for production using DNA whose sequences might later become known; since only the former method was taught by the disclosure, a claim to both methods was too broad.

---

39   See *Biogen v Medeva PLC* [1997] *supra* per the judgment of Lord Hoffmann at 50.
40   On behalf of Biogen.

While Lord Hoffmann was prepared to accept the finding of Aldous J in the Patents Court, namely, that Professor Murray's method resulted in the production of both types of antigen, he rejected a claim to monopolize *any* recombinant method of making the antigens. In the view of Lord Hoffmann, because the claims of the patent in suit were too broad, Biogen 1 did not support the invention as claimed. There was no enabling disclosure. In effect, the decision of the House suggests that a disclosure which enables the invention to deliver the goods is not necessarily an enabling invention. While such a disclosure may enable performance of the invention, it may not enable performance of all that is claimed. For Lord Hoffmann, an enabling disclosure must reflect the true extent of the claims and not merely those claims corresponding to the invention or its embodiments.

Lord Hoffmann sought to reconcile his view of disclosure with that of the Technical Board in *Biogen*. He said that the Board in that case accepted that Biogen 1 disclosed the same invention as the patent in suit, and disclosed it sufficiently to enable it to be performed to the full extent as claimed. In arriving at such a conclusion, the Board was solely concerned with whether or not both core and surface antigen was expressed as a result of the teaching in Biogen 1. Nothing was said of the possibility that the claims were too broad on the basis that the monopoly claimed extended to the production of antigen by a method not taught by Biogen 1. Biogen 1 did not teach how to produce antigen using *unsequenced* DNA, which was exactly what Biogen claimed. Once the sequences became known, it was obvious how to express antigen. In reaching its decision, the Technical Board in *Biogen* sought correspondence between the teaching in the priority document and claims, but failed to seek equivalent correspondence between the claims and teaching. When viewed from the former perspective, the technical contribution matched the claims. When viewed from the latter perspective, the claims related to methods envisaged without the invention and were, therefore, overly broad. According to Lord Hoffmann there was no difference to the issue of disclosure between the House in *Biogen* and that of the Technical Board. Therefore, there was no divergence in jurisprudence between the EPO and UK courts.

Although the reasoning of Lord Hoffmann is very persuasive, the reason for his decision is rooted in policy, namely, preserving competition. He conceded that technical contribution deserved to be recognized, but not at the expense of stifling research or healthy competition. The decision effectively set the technical contribution at naught while the effects on competition were not evaluated. Granting a monopoly in the present case, while restricting competition, would not have eliminated it, and the test, as stated, is elimination. The decision suggests that competition policy predominates and biotechnological inventions are excluded from patent protection. Other policies,[41] designed to afford patent protection, were not considered. The principle, namely, fair protection governing the requirement of sufficient disclosure, was not addressed by Lord Hoffmann. He was prepared to accept that the production of antigen was sufficiently disclosed to enable the

41   Such as the 'contract' theory.

skilled man to work the invention, yet this was not sufficient to amount to an enabling disclosure. If the 'enabling disclosure' test is more severe than 'sufficient disclosure', it may be difficult to encourage research in the future. This applies especially in pioneering technology, where broad claims provide the necessary incentive for competition in the first place. The decision of the House has provided an elegant manner in which to attack overly broad claims and in the future it may be that attacks on priority will increase.

Looking to EPO jurisprudence offered UK courts the opportunity to protect modern technology by reformulating new principles using existing policies. However, the cases demonstrate that, in the context of disclosure, such an opportunity was missed with the result that many biotechnological inventions are excluded from the scope of patent protection. If existing and clearly defined policies are not used to reformulate new principles in this manner, it is difficult to see how policies underlying the morality provision, defined less clearly, can adequately protect biotechnological inventions.

Obviousness, Inventive Step and the Man Skilled in the Art

Broad claims in respect of biotechnology often form the basis for lack of protection and have rarely found favour with courts in the UK.[42] The judiciary there has failed to address adequately the special problems posed by traditional patent law criteria for biotechnology, namely, obviousness, inventive step and the skilled man in the art. In the more traditional mechanical invention, the engineer contemplated existing art, identified the possibility of improvement, and devised a way of achieving it. Modern technology involves research projects, which by their nature can never be precisely 'devised' in advance. Research of this type is no longer carried out by a single person, however skilled, but rather by teams of people. An inventive step may lie in the smallest advance in the art, thus making non-obviousness difficult to prove. In particular, the judiciary has failed to appreciate the distinction between invention and discovery as these terms relate to biotechnology. The standard of obviousness established by the courts for biotechnological inventions is higher than that of others. The cases, discussed hereafter, demonstrate that existing substantive criteria for patentability are problematic enough and already biotechnological inventions do not receive the protection they deserve from UK courts. This suggests that, in the context of protection, additional moral constraints on the margins are undesirable.

---

42   See generally Wolfgang von Meibom and Johann Pitz, 'Broad Biotech Claims: Part 1', 1996 vol. 84 *Patent World* 29–32 at 31.

Genentech Inc.'s Patent

In *Genentech Inc.'s Patent*[43] (henceforth *Genentech*), the difficulties facing UK superior courts applying traditional patent law criteria to biotechnological inventions again become clear. The Court of Appeal concluded there was no inventive step in what Genentech did. By contrast, the EPO in Genentech did find the presence of an inventive step and, in so doing, demonstrated a willingness to protect biotechnological inventions.

Technical background   Human tissue-plasminogen activator (t-PA) is a naturally occurring protein found in blood, serum and uterine tissue.[44] t-PA activates the conversion of the precursor plasminogen into plasmin. This is an enzyme that dissolves existing blood clots by digesting fibrin, which is the protein responsible for blood coagulation. The production of t-Pa in the body is very small, but if it could be produced in quantity it would be a very desirable therapeutic agent. Cells of Bowes melanoma, a human tumor cell line, were known to secrete t-PA. However, t-PA isolated from this material was limited, and therefore the molecular cloning of the DNA sequence that encodes the protein was a worthwhile objective. By means of genetic engineering, using a particular route of recombinant DNA technology, Genentech Inc., a Californian microbiological research company, took the relevant genetic information from the cell line and expressed it in micro-organisms capable of producing t-PA as a therapeutically acceptable product. In doing this, Genentech used known recombinant DNA technology. The use of this technology to produce artificial t-Pa was not of itself a novel idea. At least four other companies were simultaneously trying to reach the same goal. The use of the technique was by no means easy. It depended upon the discovery of the DNA and amino acid sequences of t-PA.

In the UK Genentech filed 20 claims for a patent covering t-PA produced by recombinant DNA technology. Four of these were filed in May 1983, based on three American applications, with a priority date of 5 May 1982. The British patent GB-A-2119804 was issued. Immediately after its publication in 1986, the Wellcome Foundation petitioned for revocation of the patent. Shortly after, Genentech counter-petitioned and accused Wellcome of infringement.

The patent was entitled: 'Human tissue plasminogen activator, pharmaceutical compositions containing it, processes for making it, and DNA and transformed cell intermediates therefor.' The claims were drafted in broad terms as follows:

---

43   [1989] RPC 147.

44   See generally Sandra M Thomas, Keiko Kimura and Julian F Burke, 'Patenting of Recombinant Proteins: An Analysis of Tissue Plasminogen Activator (t–PA) in Europe, the United States and Japan', 1995 vol. 24 *Research Policy* 645–63 at 647.

Claim 2 was a product claim to t-PA produced by any method whatsoever.

Claim 3 was a process claim to t-PA produced by any method of recombinant DNA technology.

Because t-Pa already exists in nature, one issue for consideration was whether or not the claim related to an invention at all or was merely a discovery. Another issue was: if indeed there was an invention, did it contain an inventive step? In the Patents Court the case was heard before Whitford J,[45] who revoked the patent. The Court of Appeal confirmed the decision of Whitford J and dismissed the appeal. In the Court of Appeal the case was heard before Purchas, Mustill and Dillon L JJ.

In the Patents Court the legal basis for the decision of Whitford J lay in section 125(1) of the Patents Act 1977. Section 125(1) reads partly as follows:

For the purposes of this act an *invention for a patent* for which an application has been made or for which a patent has been granted shall, unless the context otherwise requires, be taken to be that *specified in a claim* of the specification of the application or patent, as the case may be, as interpreted by the description and any drawings contained in that specification, and the extent of the protection conferred by a patent or application for a patent shall be determined accordingly.

While Whitford J was prepared to find that what Genentech did was sufficient to amount to an invention, he did not find that it amounted to a patentable invention for the purposes of the Act. He held the claims too broad, so that section 125(1) of the Act was not satisfied, because the specification contained claims not supported by the description, and which were neither clear nor concise in that they failed to define the matter for which protection was sought. He said: 'the matter for which the applicant seeks protection is not a specification directed to a patentable invention'[46]

Whitford J revoked the patent. His decision was based on the claimed subject-matter not meeting the requirements of section 14(5)(c), namely, that the claims be supported by the description.

The Court of Appeal held that, while section 72(1)(c) of the Act permitted revocation in relation to section 14(3), it did not do so in relation to section 14(5), which was concerned with patent-granting procedure. Section 14(3) permits revocation where the teaching in the specification is not sufficient to enable the man skilled in the art to work the invention. The court said that, under the scheme of the Act, matters relating to claim width are for the Examining Officer to deal with previous to grant in accordance with the provisions of section 14(5). The Court of Appeal found that Whitford J was wrong in law to revoke the patent on

45   See *Genentech* [1987] RPC 553.
46   Ibid. at 592.

grounds of section 14(5). However, the court held that, as the invention was not an invention for the purposes of section 125(1) of the Act, it could be revoked under section 72(1)(a) of the Act, because the invention was not a patentable invention. The Court of Appeal revoked the patent because the claims were drafted in too broad a manner, that is, the extent of the monopoly claimed exceeded the technical contribution to the art.

Additionally in *Genentech*, the Court of Appeal revoked the patent on the basis that what was protected amounted to a discovery. This part of the decision was rooted in the fact that the claims were drafted in too broad a manner. Under section 1(1) of the Patents Act a patent may be granted only for an invention involving an inventive step. Section 1(2) tells that a discovery is not an invention for the purposes of the Act, but only to the extent that the invention resides in the discovery as such. Distinguishing between invention and discovery is, therefore, of crucial importance. As the *Genentech* case shows, the problem presented for the judiciary by biotechnological inventions is that confusion exists between discovery and invention.

t-PA being a naturally occurring substance, the court had first to decide whether what Genentech did was directed to an invention as required by section 1(1) of the Act, in which case it was patentable; or whether it was merely a discovery under section 1(2), in which case it was not patentable, unless it was saved by virtue of the proviso as laid down in section 1(2). It was agreed that Genentech's contribution was the discovery of the amino acid sequence of the DNA sequence of t-PA. The amino acids themselves were already known to exist, but their exact sequence was not known until Genentech completed their research. However, if the patent related to the discovery of the sequence 'as such', then it was precluded from patentability by virtue of section 1(2) of the Patents Act 1977. The issue for the court, therefore, was how to interpret the 'as such' proviso in relation to the discovery provision set out in section 1(2) of the Act.

The judgment of the court is reflected in the opinion of Purchas LJ He said:

> In my judgment the plain and ordinary interpretation to be given to the words 'only to the extent that' in conjunction with 'related to that thing as only to that such' is derived from taking the two phrases together as meaning that any of the matters listed …[s]hall not be an invention for the purposes of the Act. [a]pplying this approach to the facts of the present case, a claim for the figure 5 data[47] would be a claim to a discovery as such and extent would be disqualified by section 1(2).

For Purchas LJ, discovery as the foundation of the patent was fatal to its validity, but, he said, an applied discovery is patentable. However, the qualification was not of assistance to Genentech since Purchas LJ concluded that the invention related

---

47   This related to the information contained in the discovery, i.e., the amino acid sequencing, and constituted the invention.

to a 'discovery as such'. His reasoning was as follows. If what was discovered amounted to a new substance, then a patent could be claimed for that substance 'however made', because the normal patent law principle of absolute product protection applied. However, in the present case the discovery of the sequences of the amino acids was merely part of the process in which a product already known to exist could be manufactured. The discovery could only form the basis of an invention limited to a process patent. To allow protection against *any* use of this information (irrespective of how such information might be obtained in the future) would be tantamount to allowing protection for the discovery itself.[48] Purchas LJ concluded that the claims were 'undefined' and 'speculative'.[49] The claims, solely because they were broad, did not define any invention. In deciding that the claims were broad and speculative, Purchas LJ was unwilling to examine the difference between discovery and invention in the context of biotechnology. In biotechnological innovations the invention resides in the alteration of living processes which already exist. The reasoning of Purchas LJ suggests that innovations in respect of material already existing can never receive patent protection. Such an approach shows a profound misunderstanding of the nature of biotechnological research.

Mustill LJ was unconvinced as to whether or not any invention existed at all. He expressed the view that, because t-PA already exists in nature, it was necessary first to demonstrate that what Genentech did amounted to an invention, and then proceed to determine if the patentability criteria of novelty, inventive step and industrial applicability were satisfied.[50]

The Mustill approach suggests that naturally occurring substances were never invented. The Court of Appeal paid no attention to the fact that the molecular cloning of the DNA sequence which encoded the protein actually produced cDNA or complementary DNA, which does not itself exist in nature. The importance of this lies in the fact that cDNA does not contain 'introns', and t-PA produced by genetic engineering is, obviously, different from a sequence produced naturally. The manufacture of cDNA was a technical solution to a technical problem; it was an invention.

The Court of Appeal decision that there was no invention had clear implications for the biotechnology industry. The very nature of the industry is the manipulation of living or naturally occurring substances. The Court of Appeal decision was based on policy, namely, not permitting patent law to extend to biotechnological inventions, and is, therefore, unhelpful. The inquiry for the court was based on the premise that there was no invention. Rather than establishing whether what Genentech did amounted to an invention, the court was preoccupied with making a determination that what was achieved fell within the excluded subject-matter outlined in section 1(2) of the Act. The court was unwilling to see the invention

---

48   See *Genentech* [1989] RPC 147 at 227–8.
49   Ibid. at 228.
50   Ibid. at 263–4.

of a new product, namely cDNA. If the court acknowledged that a new product was created then the absolute product protection principle applied. In which case, according to Purchas LJ, the product 'however made' was eligible for protection. This requires extending patent protection to biotechnological processes and encourages the making of products by biotechnological means. Clearly, the Court of Appeal was reluctant to take such a step.

The Court of Appeal's treatment of obviousness and inventive step also shows reluctance to extend patent protection to biotechnological inventions as the court found that there was no inventive step in what Genentech did. By contrast, in the Patents Court, Whitford J concluded that there was an inventive step. He said:

> What Genentech did was achieved by what, in my judgment, was rather more than the exercise of proficiency; it involved laborious and costly effort and to deny any monopoly protection to those who are prepared to put as much time, skill and money into research as Genentech did is only too likely to discourage workers in this field from making advances which may be of the greatest public benefit.[51]

While the patent system was not instituted to reward labour, arguably the approach adopted by Whitford J was the correct one for biotechnological inventions. Here inventive step may lie in the smallest advance in the art, thus making non-obviousness difficult to prove. The *raison d'être* of the patent system is to encourage development in industry by rewarding the inventor with a patent in return for disclosing how to work the invention. While the decision of Whitford J is decidedly pro-patent, it nevertheless reflected public policy, namely, that a balance exists between the interests of the inventor and those of the public. The interests of the public are satisfied in preserving competition, which, although it may be significantly reduced, must not be eliminated. In biotechnological inventions the interests of the inventor are met only by rewarding the 'means' used to achieve the result; Whitford J was prepared to do this.

The Court of Appeal decision that there was no inventive step in what Genentech did was based on the fact that at least four other teams set out to do the same thing. Dillon LJ considered that the position on obviousness might have been different if no one else had set out to clone t-PA and produce human t-PA by recombinant technology. In that case, it might have been said that a novel idea for a project carried through with painstaking skill and proficiency to a successful conclusion involved an inventive step and merited patent protection. According to Dillon LJ, therefore, the invention was obvious because it was 'obvious to try'.

However, it is the judgment of Mustill LJ that is crucial because he points to the heart of the matter, namely, that the language of the Patents Act 1977 is not conducive to protecting biotechnological research. Mustill LJ saw the difficulty in arriving at a solution to the issue of obviousness as a reflection of the difficulty of

---

51   See *Genentech* [1987] RPC 553 at 596.

the subject-matter involved. For Mustill LJ, despite the fact that the technology was both novel and complex, it was not so much that the science was impenetrable, as the layman could not distinguish between techniques which were obvious and those which were not.[52] He said questions of equal difficulty arose from the need to decide what the Act meant by the terms 'inventive step' and 'obvious'; and by what characteristics were to be attributed to the 'person skilled in the art'.

According to Mustill LJ the test for inventive step must be applied to the *idea of achieving the technical solution*. 'Non-obviousness' was achieved only if the man skilled in the art cannot overcome the problem at issue without further invention on his part. Mustill LJ concluded that obstacles to Genentech could have been overcome by the skilled man with inventive capacity without further invention on his part.

It has been suggested[53] that the 'obvious to try' test applied by the court confuses knowledge and obviousness. The Court of Appeal in *Genentech* held the patent invalid for obviousness because a known goal was pursued with known techniques. Recombinant DNA technology was used to produce cDNA, a substance known to exist. In determining obviousness, the Court of Appeal (unlike the Patents Court) focused on the result obtained rather than the 'means' used to reach it. Obviousness is normally assessed by asking what would be obvious to the skilled but *unimaginative* man at the relevant date and in relation to the problem on hand. However, the approach of the Court ignored the considerable hurdles to be overcome. Such hurdles would clearly be of relevance if an assessment of the likelihood of success were part of the test. The question of which test is to be applied is a matter of policy for the courts.

Another problem confronting biotechnological inventions is: who is the skilled man in the art? In *Genentech*, Mustill LJ drew a distinction between the person skilled in the art for the purposes of section 3 and section 14(3) of the Act. The distinction went to the heart of what was one of the most confusing aspects of the case, namely, statements in the authorities seeming to deny to the skilled man any inventive capacity at all.[54] In determining who is the person 'skilled in the art', the relevant legislative provisions are as follows: 'The specification of an application shall disclose the invention in a manner which is clear enough and complete enough for the invention to be performed by a person skilled in the art' (s 14(3)) and 'An invention shall be taken to involve an inventive step if it is not obvious to a person skilled in the art, having regard to any matter which forms part of the state of the art by virtue only of section 2(2) above (and disregarding section 2(3) above)' (s 3).

---

52   See *Genentech* [1989] RPC 147 at 273.

53   See Graeme Laurie, 'Biotechnology: Facing the Problems of Patent Law', in *Innovation, Incentive and Reward*, Hume Papers on Public Policy (1997) vol. 5 46–63 at 49–50, Edinburgh University Press.

54   See *Genentech* [1989] *supra* at 279.

Mustill LJ explained that section 14(3) of the Act referred to the 'notional' addressee, that is, a person ordinarily skilled in the art, while section 3 referred to the skilled man in the particular circumstances. The particular circumstances in this case indicated that the art was difficult and skill consisted in a substantial degree of an ability to solve problems. The authorities, on the other hand, were all concerned with 'insufficiency'.[55] The language of the cases made it clear that they were concerned with the notional addressee wishing to put the invention into practice,[56] and not with the concept of 'inventive step'. According to Mustill LJ, the skilled man for section 3 purposes must possess inventive capacity, whereas the skilled man for section 14(3) purposes need not be a person of exceptional skill; he need only be capable of reading the instructions and implementing the invention.

In adopting Mustill LJ's test, namely, attributing a degree of inventiveness to the skilled man for section 3 purposes, the Court of Appeal set a higher standard for the test of obviousness in relation to biotechnological inventions. This standard discriminates against the biotechnology industry. The Court of Appeal confused the concepts of novelty and obviousness. The test of inventive step is obviousness, not novelty. In using the concepts interchangeably, the court suggests that products created by non-novel means never involve an inventive step. The Court of Appeal test, as applied, makes it difficult for products of biotechnology ever to receive patent protection.

As already noted the Court of Appeal concluded that there was no inventive step in what Genentech achieved. By contrast, the EPO in *Genentech* did find the presence of an inventive step. In denying the existence of an inventive step in *Genentech*, the Court of Appeal applied the same substantive patent law provisions as the EPO but was reluctant to extend protection to biotechnological inventions. It is the author's contention that, unless substantive patent law criteria are interpreted in a uniform manner under the EPC and the Patents Act, additional criteria relating to matters moral are undesirable.

Biogen Revisited

Inventive step was also problematic for UK courts in the *Biogen* case.[57] Here, the Patents Court and the Court of Appeal reached conflicting decisions as to whether or not there was more than one invention, while the House of Lords revoked the patent on the basis of overly broad claims.

In the Patents Court, Aldous J identified the inventive concept as relating to the expression of both core and surface antigen. The specification related to the expression of both types of antigen and it suggested use of the claimed polypeptide

---

55 *Valensi v British Radio Corp* [1973] RPC 337 outlined authorities on the issue.

56 See *Genentech* [1989] *supra* per the judgment of Mustill LJ at 280.

57 See Ian Karet, 'Priority and Sufficiency, Inventions and Obviousness', 1995 vol. 1 *European Intellectual Property Review* 42–46 at 43.

molecules in compositions and methods for the treatment and prevention of viral infections in humans. The *specification*, therefore, saw the invention as the expression of polypeptides displaying at least antigen specificity. Aldous J said: 'I conclude that there is only one invention in claim 1 when read as part of the claims and as part of the description.'[58]

The Patents Act 1977 does allow for more than one invention. Section 14(5) and section 14(6) of the Act contemplate that two or more inventions may be linked so as to form a single inventive concept. Rule 22 of the Patent Rules 1978 (made under section 14(6)) provides that claims for a product may be linked to claims for a process. While the statutory scheme clearly contemplates that a single inventive concept may give rise to more than one invention, the question of whether or not there is more than one invention cannot be answered by identifying a single inventive concept. The answer to this question is to be found in a determination of 'what it is, or what has been claimed to have been, invented'.

The Court of Appeal disagreed with Aldous J and said he did not address the question of whether the product claimed was really a group of products. By basing his decision on section 125 of the Act, Aldous J approached the question as one of documentary construction. The Court of Appeal said that, while section 125 of the Act was an appropriate place to look in order to determine whether more than one invention was specified in the claims, it could not be right to look *solely* at them. The claims were part of the description and should, therefore, be considered in the context of the specification as a whole.[59] The Court of Appeal decided to revisit the evidence upon which Aldous J made his finding of fact.[60] In doing so, the court[61] referred to *May & Baker*,[62] where the question arose as to whether or not the invention of one product was the invention of the other; unless it was, they were different inventions. Applying this test, the Court of Appeal decided that, since more than one distinct product was claimed in claim 1, the invention of one was not the invention of the other, and so, claim 1 related to more than one invention

It is difficult to reconcile the approach of the Court of Appeal in *Genentech* with *Biogen*. In *Genentech*, the Court of Appeal was reluctant to see the invention as relating to a new product. By contrast, in *Biogen* the Court of Appeal was not only willing to see the invention as relating to a new product, but rather as two inventions relating to two new products. This suggests that the court in *Genentech* was unwilling to see the presence of an invention simply because it related to a product which, it maintained, already existed in nature.

---

58   See *Biogen* [1995] RPC 25 at 43.

59   Ibid.

60   This decision was criticized by the House [1997] RPC 1; see the judgment of Lord Hoffmann at 50.

61   See *Biogen* [1995] *supra* per the judgment of Hobhouse LJ.

62   See the judgment of Lord MacDermott in *May & Baker v Boots Pure Drug Co Ltd* [1950] 67 RPC 23 at 50.

The *Biogen* case presented the courts with the opportunity to elucidate the meaning of 'inventive concept', which in the Patents Court was expressed by Aldous J as 'the idea and *decision of using expression* to obtain the polypeptides'.[63] However, the Court of Appeal disagreed with this. Hobhouse LJ opined that what Biogen did was merely to decide whether, by using standard procedures (recombinant DNA technology) in a random fashion without prior knowledge of the sequencing of the HBV, any antigen could be expressed in *E. coli*. He said: 'a mere commercial decision is not an invention'.[64]

The basis of the Court of Appeal decision lay in the fact that non-novel technology was used in the production of antigen. If so, products of recombinant DNA technology are not likely to receive patent protection in the UK. Biotechnological products almost certainly involve commercial decision-making, which, according to the Court of Appeal, does not amount to an invention. In terms of protection, the Court of Appeal decision is unhelpful, and does not deal with new problems specific to biotechnology.

By contrast, the approach to inventive concept adopted by the House of Lords is positive and helpful. Lord Hoffmann said:[65] 'The fact that a given experimental strategy was adopted for commercial reasons, because the anticipated rewards seemed to justify the necessary expenditure, is no reason why that strategy should not involve an inventive step.' For Lord Hoffmann, identifying the inventive concept was critical. In the Patents Court, Aldous J thought it was the idea of making HBV antigens by recombinant DNA technology. However, according to Lord Hoffmann, this was stating the issue in too broad a manner. A proper statement of inventive concept needed to include some express or implied reference to the problem requiring invention to overcome.[66] Therefore, mature reflection about what Aldous J intended revealed a more accurate way of stating the inventive concept as the idea of trying to express *unsequenced* eukaryotic DNA in a prokaryotic host.[67] Thus, Lord Hoffmann reformulated the inventive concept as the notion that Professor Murray's method of achieving the goal[68] would work.[69] Lord Hoffmann was prepared to *accept*, *without deciding*, that what Professor Murray did, namely, expressing random pieces of unsequenced DNA, was inventive.[70]

In reformulating what was said in the Patents Court, the House denied that the 'sweat of the brow' doctrine could operate so as to confer patent protection. Of course, it can forcibly be argued that the House is right, as labour is not protectable by patent. However, such an approach ignores the 'means' used to achieve the

---

63   See *Biogen* [1995] RPC 25, at 55.
64   Ibid. at 91.
65   See Biogen [1997] RPC 1, at 44.
66   Ibid. at 45.
67   Ibid.
68   The goal was the problem requiring invention to overcome.
69   See the judgment of the House of Lords [1997] RPC 1, per Lord Hoffmann at 43.
70   Ibid. at 46.

result. The painstaking time and effort involved in the research of new products and the special characteristics of biotechnology deserve to be recognized; product protection alone is insufficient to do this.

The Court of Appeal's reluctance to extend patent protection to biotechnological inventions is further shown by the fact that doubt existed there as to whether or not what Biogen did was patentable at all. The Court of Appeal was *prepared to hold* (although it did not actually do so) that the patent could be revoked on the ground that what was claimed did not amount to an invention. Revocation is a post-grant issue, whereas patentable subject-matter relates to pre-grant. The two issues are different. Patentable subject-matter involves inquiry into novelty, inventive step and industrial applicability. While the Court of Appeal in *Biogen* emphasized that its decision on validity did not depend on whether or not the subject-matter involved was patentable in the first instance, its *willingness* to hold that there was no invention offers little comfort to biotechnological companies who seek protection. By contrast, the House said that it was not necessary and, indeed, not helpful to inquire first whether or not there was an invention, and then proceed to determine whether the invention complied with the patentability criteria.[71] In the words of Lord Hoffmann: 'There may one day be a case in which it is necessary to decide whether something which satisfies the conditions can be called an invention, but that question can wait until it arises.'[72] In terms of protection, the approach adopted by the House is welcome and showed a willingness to examine inventive concept in the context of new technology.

Despite detailed discussion and analysis of traditional patent law criteria in *Biogen*, the concepts of disclosure and enablement remain far from settled for UK superior courts and these were (again) more recently considered by the House of Lords.

In *SmithKline Beecham plc's (Paroxetine Methanesulfonate) Patent* [2006] RPC 10, the patent in suit in an application for revocation related to the crystalline methanesulfonate salt of the known therapeutic agent paroxetine. The applicant for revocation had a prior application filed, but not published, before the priority date of the patent in suit, and asserted that the latter was invalid for lack of novelty under section 2(3) of the Patents Act 1977.

At first instance Mr J. Jacob held[73] the patent invalid. He opined there were two ways of proving anticipation, 'enabling disclosure' and 'inevitable result'; 'inevitable result' being really anticipation by happenstance. He said that the earlier proposal when carried out just happened to fall within the later claim, even though the earlier inventor had some different idea. The more central concept of 'enabling disclosure' was not a rule about 'inevitable result' but whether two inventors had in substance reached the same invention. He held the patent invalid on the ground that the earlier specification had a general disclosure of a wide variety of methods

---

71   Ibid. at 41.
72   Ibid. at 42.
73   [2003] RPC 33.

of making crystalline paroxetine methanesulfonate. This allowed the reader of the specification to overcome any problems within a reasonable time and inevitably to produce crystals described in the patent in suit.

The patentee appealed to the Court of Appeal. Here the case was heard before Aldous, Sedley and Rix L JJ. Judgment was given by Aldous LJ The Court of Appeal allowed the appeal and upheld the validity of the patent. It held that the test of an enabling disclosure involved a comparison between the claimed invention and the disclosure to decide whether in fact, not in substance, the disclosure had made the claimed invention available to the public.

The applicant for revocation appealed to the House of Lords, who allowed the appeal. Lord Hoffmann found the patent invalid for lack of novelty. Outlining the law in this area, he said there are two requirements for anticipation: prior disclosure and enablement. In respect of disclosure, Lord Hoffmann cited with approval two cases which he considered to be of unquestionable authority. The first was the judgment of Lord Westbury LC in *Hill v Evans*.[74] Here it was said:

> the antecedent statement must be such that a person of ordinary knowledge of the subject would at once perceive, understand and be able practically to apply the discovery without the necessity of making further experiments and gaining further information before the invention can be made useful.

The second passage is the judgment of the Court of Appeal (Sachs, Buckley and Orr L JJ) in *General Tire and Rubber Co. Ltd v Firestone Tyre and Rubber Co. Ltd*[75] Here it was said:

> To determine whether a patentee's claim has been anticipated by an earlier publication it is necessary to compare the earlier publication with the patentee's claim ... if the earlier publication ... discloses the same device as the device which the patentee by his claim ... asserts that he has invented, the patentee's claim has been anticipated, but not otherwise.

Lord Hoffmann then summarized the effect of these two well-known statements to be that matter relied upon as prior art must disclose subject-matter which, if performed, would necessarily result in an infringement of the patent.

In respect of enablement, according to his Lordship, that meant the ordinary skilled person would have been able to perform the invention which satisfies the requirement of disclosure. He said that the test of enablement of a prior disclosure for the purpose of anticipation was the same as the test of enablement of the patent itself for the purpose of sufficiency. He found support for this view by looking to EPO jurisprudence. He said that the Technical Board of Appeal has more than

---

74   [1862] 31 L.J. Ch (NS) 457, at 463.

75   [1972] RPC 457 at 485–6.

once held that the tests are the same[76] and he criticized the Court of Appeal for being reluctant to recognize this.

Lord Hoffmann said it was very important to keep in mind that disclosure and enablement are distinct concepts, each of which has to be satisfied and each of which has its own rules. In essence, what he said is that sufficiency of disclosure relates to devising the invention in the first place, and the skilled man is that set out in section 14(3). But an enabling disclosure relates to working the invention across the full width of what is claimed, and is tested by the skilled man as set out in section 3 of the Act.

Lord Hoffmann criticized the handling of the concepts of disclosure and enablement in the Court of Appeal, per Aldous J He said: 'The questions of disclosure and enablement are so intermingled that it is often difficult to say which of them he is talking about.'[77] As already noted, in respect of traditional patent law criteria, it is difficult to reconcile the approach of the Court of Appeal in *Genentech* with *Biogen*. However, in the early stages of interpretation of patent criteria for biotechnological inventions, such differences in approach are, perhaps, understandable. But, considering the detailed discussion that occurred by the House in *Biogen*, it is regrettable that confusion still exists as regards traditional patent law criteria in UK superior courts.

In *H. Lundbeck A/S v Generics (UK) Ltd*[78] the Court of Appeal more recently considered the issues of novelty and inventive step. In this regard, happily, conformity with existing UK jurisprudence applied. In addition, the decision also offered clarification of the sufficiency requirement as set out in *Biogen* and, it would appear, narrowed the scope of the test.

The appellant (defendant) was the proprietor of European Patent (UK) No. 034066 in respect of an antidepressant drug called Escitalopram. The respondents (claimants) brought three separate claims for revocation which were heard together in the Patents Court. The challenges to validity were founded on the prior art drug Citalopram, which was first synthesized by the appellant in 1972. Citalopram was a selective serotinin reuptake inhibitor and worked by blocking serotonin uptake by nerve cells. Citalopram was a racemate and so comprised (+) and (-) enantiomers. Escitalopram comprised the pure (+) enantiomer.

In the Patents Court, the judge rejected the grounds of attack based on lack of novelty and lack of inventive step, but held that claims 1 and 3 were invalid for insufficiency because they claimed the product made by *any* method, but the specification disclosed only two ways of making it.

In the Court of Appeal the case was heard before Lord Hoffmann, Smith and Jacob L JJ In regard to the issue of novelty, Lord Hoffmann said in order to anticipate a patent, the prior art must disclose the claimed invention and (together

---

76   T 206/83 *ICI/ Pyridine Herbicides* [1986] 5 EPOR 232, *Collaborative/ Preprorennin* [1990] EPOR 361.

77   At para. 39 of the judgment.

78   [2008] RPC 19.

with common general knowledge) enable the ordinary skilled person to perform it. He noted it was settled jurisprudence in the EPO that disclosure of a racemate does not in itself amount to disclosure of each of its enantiomers.[79] The trial judge had been correct in concluding that claim 1 was not anticipated.

In regard to the issue of obviousness, Lord Hoffmann opined that the trial judge could not have been unaware that the whole of the claimant's case on obviousness could be summed up by saying that it was *obvious to try* a particular route, and the judge correctly applied the law to the facts in this case.

In respect of sufficiency, Lord Hoffmann said that, in order to decide whether the specification is sufficient, it is first necessary to decide what the invention is. That must be found by reading and construing the claims in which the inventor identifies what he claims to be his invention. His Lordship finds support for this in the approach adopted by the Technical Board of Appeal EPO in the case of *Exxon/Fuel Oils*.[80]

Lord Hoffmann said that the trial judge founded his decision to revoke on grounds of insufficiency entirely upon the decision of the House in Biogen. However, he did not think that Biogen laid down such a broad principle. He said[81] that in Biogen:

> The House of Lords interpreted the claim as being to a class of products which satisfied the specified conditions, one of which was the molecule had been made by recombinant technology. That expression obviously includes a wide variety of possible processes. But the law of sufficiency, both in the United Kingdom and in the EPO, is that a class of products is enabled only if the skilled man can work the invention in respect of all members of the class. In my opinion, therefore, the decision in Biogen is limited to the form of claim which the House of Lords was there considering and cannot be extended to an ordinary product claim in which the product is not defined by a class of processes of manufacture.

On this basis, Lord Hoffmann allowed Lundbeck's appeal against the revocation of claims 1 and 3.

It may be that, finally, the Court of Appeal is willing to accept the policies underlying traditional patent law criteria in the UK as they apply to biotechnological inventions. On the other hand, acceptance may be illusory in that the judgment was given by Lord Hoffmann, who, it is submitted, almost always adopts a flexible approach to interpretation. Of interest, too, is that Lord Hoffmann made several references to EPO jurisprudence by which he seemed persuaded. Can it be that now patentees can rely on section 130(7) of the Act whereby certain parts are framed so as to have as nearly as practicable the same effects as their counterparts

---

79   T 296/87 (OJ 1990, 19, point 6.2), T 1048/92 and T 1046/97 Optically active trizole derivatives and compositions, point 2.12.2.

80   T 409/91 [1994] OJ EPO 653 at para. 3.3.

81   See para. 34 of the decision.

in the EPC? However, the manner in which the decision of the House in *Biogen*, regarding sufficiency of disclosure, is narrowed down may prove problematic for biotechnologists in the future.

Conclusions

Insufficient judicial guidance still exists, in particular, in UK Superior Courts on how to apply traditional substantive patent law principles so as to protect biotechnological inventions.[82] Biotechnology has presented UK courts with a new set of problems with which they are uncomfortable. Interpretation of the Patents Act 1977[83] requires striking a fine balance between the interests of the patentee and those of the public. The language of the Patents Act as understood against the background of British patent decisions is a significant factor as to why the judiciary is reluctant to uphold the validity of patents relating to new technology. While section 130(7) of the Patents Act is framed 'so as to have as nearly as practicable the same effects as their counterparts in the Convention', clearly this is not always the case, since both Genentech and Biogen were awarded patent protection in the EPO for the same inventions. The divergence in jurisprudence is difficult to explain: given the history attaching to patent law in the UK, the relative newness of the Patents Act is an unlikely reason for lack of protection; given that courts in the UK have dealt with new and complex technology before,[84] it is unlikely they are unable to cope with problems arising from biotechnology now. The patent law experience of the judiciary in UK Superior Courts to date suggests that the Patents Act 1977 will not (easily) be permitted to protect biotechnological inventions.[85] This is in stark contrast to the approach of the EPO, which showed a willingness to interpret the EPC in a flexible manner so as to afford protection to new technology.

In *Genentech* the decision of the Court of Appeal was influenced by three factors. First, t-PA is a naturally occurring substance. However, the court was not sufficiently informed in matters of science so as to appreciate what exactly Genentech did. Genentech manufactured cDNA, which differs from naturally occurring DNA in that it does not contain introns. This means that the t-PA produced by Genentech is purer than that found in nature. Had the court been willing to identify correctly the invention as relating to the production of cDNA, product protection would have ensued and other products of genetic engineering would

---

82   See Sean Hird and Michael Peeters, 'UK Protection for Recombinant DNA: Exploring the Options', 1991 vol. 9 *European Intellectual Property Review* 334–9 at 334.

83   Prior to implementation of the Patents Regulations 2000.

84   Note the 'Industrial Revolution' patents and the chemical/drug patents of the 1970s.

85   Whether or not this situation will change as a result of implementation of the Patents Regulations 2000 remains to be seen.

be eligible for patent protection in the future. However, the failure of the court to recognize the invention residing in the production of cDNA suggests that the court did not appreciate the exact nature of Genentech's research. The reluctance of the judiciary to afford protection to naturally occurring substances suggests a profound misunderstanding of the purpose and function of the biotechnology industry. Are specialized judges required?

Second, it was obvious to try to manufacture t-PA artificially. However, by focusing on the product, the court ignored the *means* by which Genentech achieved their goal. The 'obvious to try' criterion applied by the court was misdirected. The *raison d'être* of the biotechnology industry is the manipulation of living organisms. It will always be 'obvious to try' to manufacture that which occurs naturally especially where the product has beneficial qualities for mankind. The invention lay as much in the 'means' used to produce cDNA as it was in the creation of the product itself. Third, the route chosen (recombinant DNA technology) was not novel. By insisting that the route be novel, the court considered products of genetic engineering unsuitable for patent protection. As long as recombinant DNA technology remains the primary route for genetic engineering processes, 'novelty' is not satisfied.

In *Biogen* the Court of Appeal insisted that the presence of an invention must first be demonstrated before proceeding to a determination of whether or not the patentability criteria have been satisfied. This is a very strict interpretation of the language of the Patents Act. Flexibility in interpretation is required. For this reason, the approach adopted by the House of Lords in *Biogen*, that it will rarely be necessary to consider what is an invention, is welcome.

In *SmithKline* the fact that application by the Court of Appeal of substantive patent law criteria can, post *Biogen*, be subjected to criticism by the House does not auger well for biotechnologists in the future.

In *Lundbeck*, the Court of Appeal decision to narrow the test of sufficiency, rather than clarifying the law, may add uncertainty for patentees.

Given the difficulties involved in interpreting substantive patentability criteria, in particular novelty and inventive step, the ability of examiners to rule on matters moral is questionable.

Additionally, in the UK the cases indicate that policy issues dominate court decisions and have not been accommodated within the existing legislative framework of the Patents Act 1977. The scheme of the Act, together with the EPC, suggests that all inventions be treated similarly for the purposes of patent law, meaning that biotechnological inventions per se ought to be granted patents once they conform to patentability criteria. Where biotechnological inventions are concerned, the acceptability of tests relating, in particular, to disclosure is questionable. The policy of preserving competition is overemphasized and misunderstood in UK court decisions, with the result that biotechnological inventions are precluded from the scope of patent protection. The concern of competition is to ensure that, in the interests of the public, research is not stifled.

The cases demonstrate that the concepts of obviousness and inventive step, who is the man skilled in the art, and the disclosure requirement, are already sufficiently problematic for patent-granting authorities without, in addition, morality being presented as a general criterion for patentability.

Chapter 6

Protection Conferred in the United States

Introduction

The protection available in the United States to innovators in the biological field is examined in this chapter. The conclusion is that patent law policy in the US is designed to implement case law developments, suggesting there is a dynamic between the legislature, judiciary and PTO. The result is that subject-matter eligibility is determined by jurisprudence and, in this regard, the courts have taken a broad view. The flexible approach in interpreting traditional patentability criteria means that biotechnological inventions receive greater protection in the US than its European counterparts. In addition, the Plant Patent Act 1930 extends the range of subject-matter eligible for protection once plant material meets the criteria for patentability. The Plant Variety Protection Act 1970 is complementary and demonstrates the need to confer patent-like protection on a range of plant material not covered by the 1930 Act.

US Jurisprudence and Eligible Subject Matter

The statutory classes of subject-matter in the US are intended to ensure that patent protection is extended to all fields of applied technology, including plant technology. The present-day equivalent of the term 'useful arts', employed by the Founding Fathers, is 'technological arts'. Although at times difficult for patent authorities, US jurisprudence shows that interpretation of the patent laws has resulted in *all life forms* receiving protection by means of a single legal regime. By contrast, in EU Member States plant variety material receives protection by means of the Community Plant Variety Rights scheme only and animal variety material is without any form of protection (Chapter 8).

The flexible approach to statutory interpretation adopted by patent authorities in the US is illustrated in the 'products of nature' cases. A true 'product of nature' may not be patented, since it does not fit into one of the four classes of statutory subject-matter. However, patent protection may be sought for the process of *using* the newly discovered product of nature.[1] The phrase 'product of nature' has sometimes been used in a different sense, denoting unpatentability under either the

---

1   Title 35 USC s.103.

novelty or non-obviousness requirements of section 102[2] and section 103[3] of Title 35 of the United States Code (henceforth USC), respectively.

That truly natural products are not patentable is clear from an early decision of the Commissioner of Patents. In *Ex parte Latimer*[4] a claim to cellular tissue that exists in the leaves and needles of pine trees was disallowed since such fibres are not within the classes of patentable subject-matter. The basis of the decision was that fibre which nature produces is intended for the use of all men and, consequently, does not exclusively belong to any man.

More recently, in *Funk Brothers Seed Company v Kalo Inoculant Company*,[5] the Supreme Court discussed the patentability of the 'work of nature'. It was known that certain leguminous plants had the ability to 'fix' nitrogen from the air and convert it into organic nitrogenous compounds. The process depended upon the presence of bacteria infecting the root of the plant. It was also known that certain strains of bacteria inhibited the process while others did not. The applicant claimed to have invented a product composed of six strains of non-inhibitive bacteria capable of use across all leguminous plants. In delivering the opinion of the court, Mr Justice Douglas said that, because the aggregate composition of the claimed product was merely a replication of the qualities of each individual strain, there was no invention. Since the product claims did not disclose an invention within the meaning of the patent statutes, perhaps the decision is best viewed as an interpretation of the non-obviousness requirement of section 103, Title 35, USC, and not of the statutory classes of patentable subject-matter under section 101 thereof. Additionally, it seems that the claim was not for a true product of nature since the claimed culture-mixture did not exist in natural form.[6]

In *Kewanee Oil Company v Bicron Corporation et al.* (henceforth *Kewanee*),[7] the Supreme Court considered the intent of the patent laws and asserted that patent law was an evolving legal science designed to cover all emerging technologies. Chief Justice Berger, delivering the opinion of the court, said:

> The stated objective of the Constitution in granting power to Congress to legislate in the area of intellectual property is to 'promote the progress of science and useful arts'. The patent laws promote this progress by offering a right of exclusion for a limited period as an incentive for innovators to risk the often-enormous costs in

---

2   An old product derived from a new source or process is not 'new' for the purposes of meeting the novelty criterion.

3   A new product that differs from old ones only in terms of an incremental degree of purity does not meet the standard of non-obviousness.

4   1889 Comm'n 13 Dec 1889.

5   76 USPQ 280, decided 16 Feb 1948.

6   In a concurring opinion, Justice Frankfurter criticized use of terms such as 'work of nature' and 'laws of nature', holding that they were terms infected with too much ambiguity and equivocation.

7   181 USPQ 673, decided 13 May 1974.

terms of time, research and development. The productive effort thereby fostered will have a positive effect on society through the introduction of new products and processes of manufacture into the economy, and the emanations by way of increased employment and better lives for citizens.

Several points are worthy of note. First, the progress of science is a prerequisite to the grant of a temporary monopoly. Second, while the word 'progress' is not defined, it clearly is equated with new products of manufacture, new processes of manufacture, increased employment and, ultimately, better lives for citizens. Third, while Congress is empowered to legislate to achieve the aims, it is under no positive obligation to do so.

The Supreme Court in *Kewanee* highlighted that constitutional policy considerations underlie patent law and continue to apply. Congress has always responded positively to the bidding of the Constitution and implemented policy in this regard, evidenced by the fact that the Plant Patent Act 1930 was enacted to give protection to breeders of asexually reproduced plant varieties. In this manner, working together, the judiciary and legislature ensure that new technologies are protected. By contrast, the underlying policy of Article 53 of the European Patent Convention (henceforth EPC) is, or has become, unclear.[8] Additionally, the EPO is both policymaker and enforcer, thus exercising a degree of control in respect of patentable subject-matter not originally intended under the EPC.

In *Re Merat and Cochez*[9] the issue of patentable subject-matter arose. Here the claim related to the discovery of 'dwarfism genes' in chickens which, when employed in controlled breeding processes, produced dwarf hens, which could then be mated with normal roosters to produce chickens with a normal body weight. The Board of Appeals of the United States Patent and Trademark Office (henceforth PTO) held that the subject-matter was not patentable on the grounds that a method of breeding animals was not a 'process' and that a thing occurring in nature (the chicken) under controlled propagation was not a 'manufacture' within the meaning of section 101, Title 35, USC. On appeal, the Court of Customs and Patent Appeals (henceforth CCPA) held only that the invention was not described in a precise and essentially reproducible manner as required by section 112, Title 35, USC. The issue of whether or not living matter inventions could be patented, unresolved in Merat and Cochez, was unsatisfactory.

In *Re Bergy, Coats and Malik*[10] (henceforth *Bergy*) the CCPA considered the patentability of living organisms. A claim involving biologically pure culture and man-made, not existing in nature, was in issue. The Board of Appeals of the PTO rejected the claim on the basis that section 101, defining the categories of patentable subject-matter, did not extend to living organisms. The CCPA reversed, noting

---

8   See generally Ch. 4 heretofore.
9   186 USPQ 471, decided 7 Aug 1975.
10 195 USPQ 344, decided 6 Oct 1977.

that the problem was one of first impression.[11] The CCPA reasoned that the claim was not to a 'product of nature', since the culture did not exist in nature in pure form, and was produced only under carefully controlled laboratory conditions. Additionally, the CCPA noted that prior cases permitted *process* claims to methods employing living organisms, and here, the claimed product, though living, was '[a]n industrial product used in an industrial process, a useful or technological art if ever there was one'.[12]

The decision is based on logic and legal practice. Because patents are permitted for *processes* with 'life' in them, by analogy patents should not be refused for *products* with 'life' in them. The CCPA concluded that it was in the public interest to include micro-organisms within the terms 'manufacture' and 'composition of matter'.[13] The pragmatic approach adopted by the court ensures that biotechnological inventions receive patent protection. However, the effect of the case is limited because the CCPA noted that it was not deciding whether living things in general, or, at most, whether any living things other than micro-organisms, are within section 101.[14] By contrast, the approach under the EPC is that micro-organisms are classified as an exception to the exclusion provision outlined in Article 53(b).

The caveat outlined in *Bergy* suggests that some uncertainty persisted in respect of what constituted patentable subject-matter for the purposes of section 101. Therefore, it was imperative that the Supreme Court rule on the matter. The opportunity presented itself in 1980.

In *Diamond, Commissioner of Patents and Trademarks v Chakrabarty*[15] (henceforth *Chakrabarty*) the issue of whether or not biological processes could be patented was finally resolved. The applicant used the technique of 'genetic engineering' to create a new strain of bacteria with improved capacity for degrading crude oil. The claim related to a bacterium containing at least two stable energy-generating plasmids, each of which provided a separate hydrocarbon pathway. The Board of Appeal's rejection of the application was based on the living nature of the claimed product. The CCPA reversed the decision of the Board and there was an appeal to the Supreme Court, which held that genetically altered living organisms constituted patentable subject-matter under section 101, Title 35, USC. The *Chakrabarty* decision is important in appreciating the Supreme Court's emphasis on limiting judicially created exceptions to patentable subject-matter under section 101. In delivering the opinion of the court, Chief Justice Berger noted that the question was 'a narrow one of statutory interpretation: whether or not micro-organisms constituted a manufacture or composition of matter'.

---

11   Ibid. at Point 4 of the Opinion.
12   Ibid. at 7.
13   Ibid. at 4.
14   Ibid.
15   206 USPQ 193, decided 16 June 1980.

In *Chakrabarty* the Supreme Court reached several conclusions, summarized as follows.[16]

- The plain meaning of the statutory language was indicative of Congress's intent that the patent laws should be given wide scope. The terms 'manufacture' and 'composition of matter' were broad in themselves, even before qualification of the expansive term 'any'.
- The legislative history of the Patent Act 1952 made it clear that Congress intended a broad construction. Referring to committee reports accompanying the recodification of the 1952 Act, the Chief Justice noted that Congress intended statutory subject-matter to be all-inclusive.
- While laws of nature, physical phenomena and abstract ideas are not patentable per se, to the extent that they can be applied to a new and useful end *re* patentable. Because the claimed bacterium was a product of human ingenuity, having a distinct name, character and use, it was patentable.
- Neither the passing of the Plant Patent Act 1930 nor the Plant Variety Protection Act 1970 evidenced an understanding by Congress that the terms 'manufacture' and 'composition of matter' did not include living things. The language of the House and Senate Committee reports suggested there was a clear and logical distinction between the discovery of a new variety of plant and of certain inanimate things.[17]
- For the court the relevant distinction was not between living and inanimate things, but between products of nature, whether living or not, and human-made inventions. Because the origin of patent law in the US is constitutional, Congress employed broad general language in patent Acts ensuring that technological inventions often unforeseeable receive protection.
- The fact that genetic technology was unforeseen at the time when Congress enacted the patent Acts did not lead to the conclusion that such technology could not receive protection until and unless Congress expressly so authorized.
- Arguments against the technology on hazard grounds should be addressed to Congress and the Executive, not the judiciary.

The Supreme Court refused to consider the 'grave risks' associated with genetic research, preferring the view that grant or denial of patents is not likely to put an end to the research or its attendant risks. The court said:

---

16 See *New Developments in Biotechnology*, Office of Technology Assessment Special Report no. 5, US Congress (Apr 1989) Washington, DC.

17 It was suggested by the Commissioner in argument that by enacting legislation which conferred protection to living matter in the form of plants Congress did not intend such terms to encompass living matter in animal form.

> We are without competence to entertain these arguments ... [t]he choice we
> are urged to make is of high policy for resolution within the legislative process
> after the kind of investigation, examination, and study that legislative bodies can
> provide and courts cannot.[18]

This suggests there is no attempt in the US to deny patents on the basis of moral concerns.

In the wake of the *Chakrabarty* decision, in 1984 the PTO announced a policy of 'pre-emption'. This meant that, if subject-matter was patentable by virtue of the Plant Patent Act 1930 or protectable under the Plant Variety Protection Act 1970, it was pre-empted from patentability under section 101 of the Utility Patent Act. The policy was in direct conflict with the decision of the Supreme Court in *Chakrabarty*. However, as the *Chakrabarty* case dealt only with genetically altered bacteria, not covered by the plant-specific statutes, such a policy was prima facie legitimate, and would have the effect of reducing the workload of the PTO.[19] Therefore, whether or not plant-specific statutes were the exclusive forms of protection for plants remained unclear.

In *Ex parte Hibberd*[20] (henceforth *Hibberd*) the validity of the PTO pre-emption policy was tested. Here the applicants developed maize plant technology that produced corn having an increased level of tryptophan.[21] Such corn is considered as having greater nutritional value than ordinary corn. The invention resided in the production of an enzyme, AS, which counteracted the inhibitory effects of tryptophan in plants in which it was overproduced. The examiner took the position that subject-matter potentially protectable under the plant-specific statutes was not eligible for protection under the general utility patent statute. However, the Board of Patent Appeals and Interferences[22] PTO disagreed and held that plants, plant seeds and plant-tissue cultures constitute patentable subject-matter within the meaning of section 101. In reaching its decision, the Board looked to the legislative history of the two plant Acts[23] which indicated an absence of congressional intent to narrow the scope of protection otherwise available under section 101. On the contrary, the Board noted the intent of Congress was to remove two perceived obstacles to patenting plants, namely, the 'product of nature' doctrine and the inability to provide an adequately enabling description. By contrast, in reaching

---

18   206 USPQ 193 at Point 11 of the Opinion.
19   The Plant Variety Protection Act is not administered by the EPO.
20   227 USPQ 443, decided 24 Sept 1985.
21   An enzyme responsible for growth in corn plants.
22   An interference is a proceeding instituted for the purpose of determining the question of priority and patentability of invention between two or more parties claiming substantially the same invention. The PTO may declare an interference when one patent application claims substantially the same patentable invention as is claimed in one or more other applications or issued patents.
23   Plant Patent Act 1930 and the Plant Variety Protection Act 1970.

decisions the EPO rarely looks to the legislative history of the EPC and does not adopt a flexible enough approach to interpretation. The value of *Hibberd* lies in the fact that the pre-emption doctrine was laid to rest and the way was now open to protecting *all* life forms[24] under a single regime.

In *Ex parte Allen*[25] (henceforth *Allen*) the patenting of multicellular animals was in issue. Here the applicants developed a method of inducing polyploidy in Pacific oysters, the effect of which was to increase growth. The advantage of producing sterile oysters is that they do not devote significant portions of body weight to reproduction and, therefore, remain edible all year round. The examiner allowed certain *process* claims but refused the applicants' *product-by-process* claims on the grounds that the claimed subject-matter was directed to 'living entities' and, therefore, not patentable under section 101. The Board of Appeals disagreed and reversed the decision. The Board took the view that, although 'laws of nature' control breeding processes, this did not address the issue of whether the subject-matter is a non-naturally occurring manufacture or composition of matter.[26] However, it should be noted that the Board did affirm rejection of the claims on the grounds of non-obviousness under section 103, Title 35, USC.

As a result of the decision in *Allen*, and to help implement case law developments in the PTO Rules, the Commissioner of Patents and Trademarks announced a new policy in relation to patentable subject-matter under section 101. The policy, announced on 7 April 1987, reads partly as follows: 'The PTO now considers non-naturally occurring, non-human, multicellular living organisms, including animals, to be patentable subject-matter within the scope of 35 USC 101.'

The new policy offers protection for all biotechnological inventions, including plant technology, under a single legal regime, and is welcome. Now, genetically engineered plant varieties are eligible for utility patent protection. Additionally, the policy paved the way for patenting animals. In this regard, it was not long before the PTO had to deal with an application concerning not merely an animal, but a transgenic animal. On 13 April 1988 Harvard University was granted a patent covering a new breed of mouse produced by genetic alteration.[27] As already noted,[28] the mouse was genetically engineered to be susceptible to cancer, thus facilitating cancer therapy research. While the decision in *Onco-Mouse* was welcome, the case raised certain issues requiring clarification. The insertion of a human-activated oncogene sequence into a non-human animal necessarily raises questions as to the dividing line between non-human and human life. The case resulted in public debate on scientific, regulatory and economic issues. The decision in *Chakrabarty*, coupled with subsequent action

---

24   However, as the decision is by an administrative body within the PTO and relates to an issue of law, it will not bind an Article 111 Constitutional Court that is called upon to decide that issue.

25   2 USPQ 1425, decided 3 Apr 1987.

26   Ibid. at 1426–7.

27   Known as Onco-Mouse.

28   See generally Ch. 4 heretofore.

taken by Congress and the executive, provided great economic stimulus to patenting micro-organisms and has been responsible for the growth of the biotechnology industry. Revisions in federal patent policy were followed by legislative reform. The Patent and Trademark Amendments Act 1980[29] encouraged the patenting and commercialization of government-funded inventions by allowing small businesses and non-profit-making organizations to retain ownership of inventions developed in the course of such research. In 1983 the policy was extended to larger businesses by Executive Order. The Technology Transfer Act 1986[30] granted federal authority to form consortia with private concerns (thereby obviating the need to conform to anti-trust law). In 1987 an Executive Order further encouraged the transfer of technology by allowing patent rights to be transferred to government grantees. In tandem with legislative reform, the judiciary embarked on judicial legislation in respect of developments in biotechnology, and, arguably, now patentees know when their actions are safe and lawful.

The intent of the patent laws in the US suggests drawing upon the latest state of scientific knowledge to interpret patentability criteria not defined more precisely by the legislature and which, by their nature, require judicial determination. Because the origin of patent law is constitutional, Congress employed broad general language in the patents Acts and the Supreme Court limited judicially created exceptions to patentable subject-matter. In this manner, technological inventions, often unforeseeable, receive protection.

Protection under the Plant Patent Act 1930

The general purpose in having plant protection is to afford agriculture, so far as practicable, the same opportunity to participate in the benefits of the patent system as has been given industry.[31] In 1930 Congress provided for the first time for patents for plants[32] on the basis that such technology meets traditional patentability criteria. However, in this regard, due to the nature of plant technology a fresh approach to legislation was required. The result was the Plant Patent Act 1930 (henceforth PPA) and flexibility is seen in:

- eligibility for protection
- scope of exclusive rights
- description requirements
- requirements for patentability.

---

29   As amended in 1984, Public Law 96-620.

30   Public Law 99-502.

31   The House and Senate Committee Reports outline the purpose of plant patents; see Senate Report no. 315, 71st. Congress, 2d session, 1930, and House of Representatives Report 1129, 71st Congress, 2d session, 1930.

32   Act of 23 May 1930, ch. 312, 46 Stat 376.

The Act amended section 4886 of the Revised Statutes[33] to include persons who have 'invented or discovered and asexually reproduced any distinct and new variety of plant, other than a tuber propagated plant'.[34] It amended section 4884[35] to include the phrase 'in the case of a plant patent the exclusive right to reproduce the plant'. It amended section 4888[36] to include the provision 'no plant patent shall be declared invalid on the ground of non-compliance with this section if the description is made as complete as is reasonably possible'. The Act was limited to inventions involving asexual reproduction, that is, from cuttings, budding or grafting, because at the time of its enactment it was thought that only asexual plants could breed 'true to type'.

In the Patent Act 1952 the plant patent provisions were placed in separate sections of Title 35, USC, namely, sections 161–4. Section 162 added to the completeness of description provision, namely, 'the claim in the specification shall be in formal terms to the plant shown and described'. Thus, as for utility patents, the description must be 'enabling'. Section 163 defined the exclusive rights granted by a plant patent as 'the right to exclude others from asexually reproducing the plant or selling or using the plant so reproduced'. Therefore, protection is limited to sale or use of the entire plant. In 1954 Congress amended section 161[37] to increase the classes of plant eligible for patent protection. Protection now extended to 'a distinct and new variety of plant, including cultivated sports, mutants, hybrids and newly found seedling'. The purpose of the amendment was to indicate clearly that plant seedlings discovered and asexually reproduced having new characteristics distinct from other known plants are patentable.[38] In this manner, Congress increased subject-matter patentability in the plant variety area. By contrast, it is suggested, the EPO has adopted a very narrow approach to subject-matter patentability.

Requirements for Patentability

Under the PPA a new plant variety is patentable if it is distinct and asexually reproducible and meets the general requirements of novelty, originality and non-obviousness. In this regard, a flexible approach is adopted and judicial decisions show that plant varieties meet patentability criteria.

That the term 'plant' in section 161, Title 35, USC, covers plants in their ordinary and common meaning rather than the strict scientific sense is seen from

---

33 On the classes of eligible subject-matter.

34 This exception is made because tuber-propagated plants, unique among asexually reproduced plants, is propagated by the same part of the plant that is sold as food.

35 On the grant of exclusive rights.

36 On full disclosure in the specification.

37 Act of 3 Sept 1954, Pub L no. 83-775, 68 Stat 1190.

38 See Senate Report 1937, 83d Congress 2d session (1954); and HR Report 1445, 83d Congress 2d session (1954).

examination of the decision in *Re Arzberger*.[39] Here, the applicants sought to claim a species of bacteria cultured from Louisiana cane field soil. Reproduction was asexual by binary fusion. The species was useful for producing butyl alcohol, acetone and ethyl alcohol. The CCPA affirmed the decision of the Board of Appeals PTO, rejecting the claim on the grounds that the subject-matter did not constitute a plant. The CCPA conceded that, while bacteria 'are usually scientifically classified as plants', the legislative history of the Act made it clear that 'plant' was used in its popular sense.

Two points are worthy of note. First, the court was willing to look to the legislative history of the Act in order to determine its intent, namely, removal of existing discrimination between plant inventors and industrial developers and stimulating capital investment in plant breeding with the aim of promoting development through private funds. By contrast, in determining *intent* the European Patent Office rarely looks to the legislative history of the EPC, suggesting that the purpose of excluding patent protection for plant varieties is to defer to the provisions of the International Union for the Protection of New Varieties of Plants. Act (UPOV). However, in the light of the revised UPOV 1991, such purpose is now eliminated and there seems to be no valid reason why the ban in Article 53 EPC should be retained.[40] Second, the court saw 'privatization' as a necessary incentive in the development of agriculture. The pragmatic approach adopted by the court highlights the *raison d'être* of the patent system and gives meaning to the policy that patents should be available for inventions in any type of industry, including agriculture. In contrast, the EPO by excluding plant varieties from patent protection continues 'discrimination' between plant inventors and industrial developers with the result that in European states plant breeding and research will, most likely, continue to rely on government funding and public institutions.

Development of a new variety of asexually reproduced plant involves three major steps, namely, cultivation or discovery, identification and asexual reproduction.[41] Cultivation or discovery is not an element of the act of invention of a new plant variety.[42] In *Ex parte Moore*[43] the Board of Appeals PTO said: 'although one may find a plant, he has not discovered a new variety if he has no appreciation that the plant is a distinct and new variety'. Therefore, simply to become aware of the existence of a plant without appreciation that it is a new variety is not sufficient to entitle the finder to patentability. Identification of the characteristics that establish the variety as new and distinct is a necessary element of the act of invention. In this sense, the novelty requirement for plant patents overlaps the requirement of distinctness. The alleged new variety must be distinct in terms of

---

39   46 USPQ 32, decided 24 June 1940.

40   The 'bar' on double protection no longer exists under UPOV 1991.

41   See generally Donald S Chisum, *Patents: A Treatise on the Law of Patentability, Validity and Infringement* (1978) Matthew Bender and Co.

42   Ibid.

43   115 USPQ 145 [1957].

its characteristics from plants in the 'prior art'.[44] The rationale of Congress was that discovery of a new plant from cultivation is unique, and could not be repeated by nature, nor reproduced by nature unaided by man. In this regard, discovery amounts to invention.

That plant varieties are more freely patentable than other types of invention is seen from examination of the statutory bar provisions of section 102, Title 35, USC, relating to novelty. Section 102 reads as follows:

> A person shall be entitled to a patent unless:
>
> (b) the invention was patented or described in a printed publication in this or a foreign country or in public use or on sale in this country, more than one year prior to the date of the application for patent in the United States.

A couple of points are worthy of note. First, publication of an 'enabling description' is required before the statutory bar comes into effect. However, no matter how complete, a written description of a new variety will normally not enable one with skill in the art to reproduce the variety exactly. Second, while Congress made an exception as to disclosure in the specification of a plant (the 'best mode' of performing the invention need not be disclosed; rather, the description must be as complete as is reasonably possible), it did not do so as to publication. This suggests that publications of plant varieties more than one year prior to patent application will rarely pose a bar to patentability.

Whether or not the requirement of non-obviousness for utility patents applies with equal force to plant patents is unclear. Section 161, Title 35, USC provides that 'the provisions of this title relating to patents for inventions shall apply to plant patents, except as otherwise provided'. Such other provisions could include section 103, containing the judicially created requirement of 'invention'. However, the legislative history of the PPA makes no mention of the invention requirement and seems to regard distinctness and asexual reproduction as the substantive requirements for plant patents.[45] Support for this is found in the reports themselves which state that 'it is immaterial whether in the judgment of the patent office the new characteristics are *inferior* or superior to those of existing varieties'.[46] Yet, a different approach was taken in *Yoder Bros. Inc. v California-Florida Plant Corp*[47] (henceforth *Yoder*), and it was held[48] that the non-obvious requirement did apply to plant patents. The court said: 'We think that the most promising approach toward the obvious requirement for plant patents is the reference to the underlying

---

44   Title 35 USC s.102(e).

45   See generally Donald S Chisum, *Patents: A Treatise on the Law of Patentability, Validity and Infringement* (1978), *supra*.

46   Ibid.

47   193 USPQ 264 [1976].

48   By the 5th Circuit.

constitutional standard that it codifies, namely, invention'. The rationale of the court was that the thrust of the 'invention' requirement is to ensure that minor improvements do not receive monopoly protection. Nevertheless, the court in *Yoder* proceeded to hold that developing new varieties retaining desirable qualities of the parent stock and adding significant improvements through asexual reproduction was no minor improvement and was sufficient to constitute 'invention' for the purposes of the PPA.

The utility patent system in the US has given industry there the opportunity to participate in benefits otherwise not available. The PPA offers breeders the same opportunity as industrialists in relation to agricultural and horticultural processes. There is no longer discrimination in the US between inventors of plant technology and industrial developers both receive protection by means of patent. Plant patents are available on the basis that such technology meets patentability requirements as interpreted by the judiciary. By contrast, under the EPC the patent system discriminates between plant breeders and industrialists and a dual system of protection operates. What is required is a single regime whereby material meeting patentability criteria receives protection.

Protection under the Plant Variety Protection Act 1970

Scientifically, once it became known that plants reproducing in a sexual manner breed 'true to type', the legislature in the US enacted legislation to protect plant variety material. In 1970 Congress enacted the Plant Variety Protection Act (henceforth PVPA).[49] The purpose of the PVPA, succinctly outlined in the Preamble, is:

> [t]o encourage the development of novel varieties of sexually reproduced plants and to make them available to the public, providing protection available to those who breed, develop, or discover them, and thereby promoting progress in agriculture in the public interest.

Congress expressly provided that it was relying on both the patent-copyright clause and the interstate commerce clause[50] of Article 1 of the Constitution as the source of authority for the PVPA. Therefore, the legal basis of the PVPA is the same as for the patent laws. This highlights the similarity between the two systems of protection, a theme taken up by the courts. In *Public Varieties of Mississippi Inc. v Sun Valley Seed Co. Inc.*[51] the court said: 'Because of the similarity in purpose and

---

49   82 Statute 1542, 7 USC s.2321 *et seq.*, Pub L 91-557.
50   Clause 3 reads: 'Congress shall have power to regulate commerce with foreign nations, and among the several states, and with the Indian tribes'.
51   14 USPQ 2d 2005, [1990].

construction between the PVPA and the patent laws, cases construing the patent statutes supply compelling analogies to aid the court in interpreting the PVPA'.

Plant variety protection certificates are issued by the Secretary of Agriculture after examination by the Plant Variety Protection Office. The certification standards of the PVPA are less rigorous than for utility and plant patents and the two systems differ in terms of the scope of protection conferred. To qualify for protection under the PVPA, the variety must be 'new, distinct, uniform and stable'. Similarities between the two systems include statutory bar provisions and a 'first to breed' approach along the lines of the 'first to invent' for patents. A protection certificate confers on the holder the right to exclude others 'from selling the variety, or offering it for sale, or reproducing it, or importing it, or exporting it, or using it in producing a hybrid or different variety therefrom'.

Worthy of note are three differences between plant variety and patent protection in terms of the scope of the exclusive rights conferred.[52] First, the term of a certificate may be shortened if it does not issue within three years of filing and delay is attributable to the applicant.[53] Second, the administrator of the PVPA is entitled to issue a compulsory licence when it is necessary to ensure an adequate supply of food in the US.[54] Third, the owner of a certificate may obtain rights prior to issuance by circulating seed to the public with a proprietary notice. After issuance, the owner may then collect infringement damages for unauthorized acts occurring prior to issuance.

Under the PVPA, while seed-bearing plants are certifiable, fungi, bacteria and first-generation hybrids are excluded. Although fungi and bacteria are sometimes taxonomically classified as plants, they are not so considered for the purposes of the Act. First-generation hybrids are excluded because they are inherently genetically unstable (resulting from cross-fertilization) and are considered incapable of reproducing with uniform characteristics. In 1994 Congress enacted changes to the PVPA in order to enable the US to ratify the revised UPOV 1991,[55] so that now plant variety legislation protects hybrids but continues the exclusion of both fungi and bacteria.

Section 2542 *et seq.*[56] provide exemptions[57] for acts that might constitute infringements. The underlying purpose of the exemptions is to respect the rights

---

52   See generally Donald S Chisum, *Patents: A Treatise on the Law of Patentability, Validity and Infringement* (1978), *supra*.

53   7 USC s.2483(a).

54   Ibid. at s.2404.

55   1991 Act Art. 3 UPOV requires protection for all plant genera and species.

56   To s.2545.

57   Section 2542 permits a person to reproduce and sell a variety if he has developed that variety more than one year before the effective filing date of an adverse application. Section 2543 permits farmers to sell surplus seed to other farmers so long as neither seller nor buyer is primarily engaged in producing the seed for sale. Section 2544 and s.2545 allow for the use of protected varieties in research.

of growers in the context of the realities of farm life, as well as society's right to have the largest possible gene pool available for plant breeding research. In this regard, whether or not a sufficient balance exists is uncertain.

Whether it is best to obtain protection for new plant material under the PPA 1930, PVPA 1970 or utility patent statutes depends on the circumstances. The areas of subject-matter protected by the three statutes overlap. Attempts to obtain dual protection may be objected to on double patenting grounds. If a choice between the three options must be made, the following are deserving of consideration:[58]

- The PPA limits infringement to acts of asexual reproduction. Neither the PVPA or utility patent statutes contain any such limitation.
- The utility patent is subject to a limited judicially defined experimental use exception, whereas in this regard the PVPA is subject to an express statutory provision.
- A utility patent and the PVPA permit the applicant to claim multiple parts of the plant, including plant genomes coding for non-plant proteins. Under the PPA, protection is limited to sale or use of the entire plant.
- A utility patent may be used to claim multiple varieties. Such is not the case for either the PPA or the PVPA.
- The non-obvious requirement applies to both the utility patent and the PPA. By contrast, the PVPA requires only that a plant meet the required standards of a 'novel' variety, namely, distinctness, uniformity and stability. However, while a utility patent may offer a greater scope of protection, a particular plant variety may not satisfy the non-obvious requirement and may be forced to seek protection under the PVPA.

Protection under the PVPA 1970 is akin to patent protection. It is recognized in the US that variety protection is required for promoting development of agriculture in the public interest. In respect of plant technology, the legislature enacted legislation to meet new developments and the judiciary responded.

Conclusions

The policy underlying patent law in the US is constitutional and continues to operate. The policy is designed to encourage case law developments. In this regard, there is a dynamic between the legislature, judiciary and PTO, absent in European *fora*. The result is that biotechnological inventions in the US receive greater protection than in European states under the EPC or the Directive on the Legal Protection of Biotechnological Inventions 1998.

---

58   See generally Donald S Chisum, *Patents: A Treatise on the Law of Patentability, Validity and Infringement* (1978), *supra*.

The mere existence of the Plant Patent Act 1930 in the US shows, first, that plant material does meet the criteria for patentability, and, second, the extensive range of subject-matter eligible for protection. The existence of the Plant Variety Protection Act 1970 is complementary and demonstrates a need to confer patent-like protection on a range of plant material not covered by the Plant Patent Act.

The flexible approach of the judiciary to interpreting traditional patentability criteria, without difficult assessment of additional marginal constraints, suggests that in the US there exists the political will to adequately protect biotechnological inventions.

Chapter 7
Protection Conferred under
the Biotechnology Directive 1998

Introduction

Most people are now aware of the cultural, societal, economic and political changes brought about by the life sciences and information technology. The so-called 'information society' depends on its ability and capacity to produce, distribute and exploit knowledge.[1] The application of modern biotechnology is already having a considerable impact on the health care, agriculture and environmental sectors. At the same time the way in which people exchange ideas and information is being revolutionized through the influence of communication technologies.[2] Against this background, it is understandable that many countries are addressing moral issues to help decision-makers to assess the impact of these technologies on society and to head off any harmful developments.

Biotechnology, through its wide-ranging implications especially concerning health care, agriculture and the environment, as well as through the new knowledge it offers, will have a considerable impact on our way of life. Biotechnology also offers specific new possibilities for information and interventions affecting human life[3] and raising basic moral values. For these reasons biotechnology attracts considerable public interest and debate, some of it confused. This confusion impacts on industry[4] and can adversely influence the climate for innovation and development of biotechnology.[5] Nonetheless the questions arising in public debate remain and include:

---

1   See generally Christian Lenk, Nils Hoppe and Roberto Andorno (eds), *Ethics and Law of Intellectual Property* (2007) Ashgate Publishing, UK.

2   See Ms Noelle Lenoir, President, European Group on Ethics, in a report to the European Commission, *The European Group on Ethics in Science and New Technologies*, Dec 1999 2–3 (www.europa.eu.int/comm/sg/ethics).

3   See Nuffield Council on Bioethics report 'Ethical Principles: Respect for Human Lives and the Human Body', in *Human Tissue Ethical and Legal Issues* (April 1995) 39–54 at 39.

4   See Nigel Jones, 'The New Draft Biotechnology Directive', 1996 vol. 6 *European Intellectual Property Review* 363–5 at 364.

5   See the European Group on Ethics in Science and New Technologies' Commission Communication, *Promoting the Competitive Environment for the Industrial Activities Based on Biotechnology within the Community*, Dec 1999, SEC (91) 629 final.

- moral considerations relating to human life or in research on the human genome;
- animal welfare issues;
- issues relating to the limits of intellectual property rights;
- environmental as well as health and safety issues.

All of these concerns are important, and both national and Community policymakers must ensure that legislative measures respond to meet these needs. The politicians of the European Union (henceforth EU) have spent a decade in debating, and finally agreeing, the terms of the Directive on the Legal Protection of Biotechnological Inventions 1998 (henceforth the Directive).[6] The Directive aims to harmonize the criteria for the patentability of genetic material across the EU Member States and to facilitate a uniform application of the immorality exclusion in all Member States, so as to encourage investment in biotechnology.[7] A consequence of this is that patent law has been transformed from the obscure province of specialist practitioners to a focus for public debate.

Although morality has been forced back onto the legal agenda, the legislative history of the Directive demonstrates that the relationship between the two has been uneasy.[8]

In 1985 the European Commission White Paper[9] on completing the internal market stated that differences in intellectual property laws among Member States had a direct and negative impact on intra-Community trade. This prompted a proposal for a Directive from the Commission in relation to patenting of biotechnological inventions. The original proposal for a Directive was adopted by the European Commission in October 1988.[10] However, the procedure for its adoption was slowed down by, primarily, moral issues regarding the patentability of living matter, as the proposal did not take into account sufficiently this dimension of patenting biotechnological inventions. Negotiations began between the Commission and the European Parliament to modify the proposal to include some basic moral principles. In effect, the European Parliament demanded an 'integrated approach', that is, the inclusion of moral, social and legal aspects into the intellectual property programme.[11] In particular, the European Parliament wanted principles reflecting

6   See the *European Group on Ethics in Science and New Technologies* report, *supra* at 65.

7   See Julia Black, 'Regulation as Facilitation: Negotiating the Genetic Revol.ution', 1998 vol. 61 *Modern Law Review* 621–60 at 645.

8   See Lionel Bently and Brad Sherman, 'The Ethics of Patenting: Towards a Transgenic Patent System', 1995 vol. 3 *Medical Law Review*, 275–91 at 277.

9   *Growth, Competitiveness and Employment*, 5 Dec 1993 COM (93) 700 final.

10   COM (88) 496 final.

11   See Sigrid Srerckx, 'European Patent Law and Biotechnological Inventions', in Sigrid Sterckx (ed.), *Biotechnology, Patents and Morality* (1997) 1–55 at 20, Ashgate Publishing, UK.

derogation for farmers, and exclusion from patentability of the human body and its parts, and of inventions contrary to *ordre public* or morality.[12]

The negotiations between the Commission and Parliament lasted four years.[13] In February 1994 the Council announced its position. The differences between the Council and Parliament resulted in a Conciliation Committee[14] agreeing a joint text.[15] However, on 1 March 1995 the joint text was rejected by the European Parliament.[16] A new Commission proposal was then drawn up[17] in the light of the reasons put forward by Parliament, with a view to finding the best solution in this highly sensitive field.

For the Commission, without common legislation compatible with the single market, European research was at a disadvantage compared with competitors in third countries. The legal protection of biotechnological inventions needed to be clarified. Whereas current 'European' patent law, namely, Strasbourg and European Patent Convention (henceforth EPC) law, was drafted in the 1960s and 1970s, at a time when the scope offered by biotechnology could not be imagined, all the discussions that took place regarding the initial proposal in 1988 confirmed the need for the Member States laws to be further harmonized. It was agreed that the legal protection of biotechnological material must be founded on the basic principles of existing patent law. But to what extent? For EPC Contracting States, except for exclusions to patentability outlined in Article 53, biotechnological material was patentable under patent law as it then stood. Thus, patentability of biotechnological material was not something introduced for the first time by the Directive. On the one hand, opponents[18] of the legal protection of biotechnological inventions hold the view that no material occurring in nature can be subjected to patents. However, on the other hand, it has been argued elsewhere the Directive follows a well-balanced approach and, in principle, it imposes no specific restrictions on the patentability of human material.[19]

The proposed Directive included principles of non-commercialization of the human body and the free and informed consent of the person from whom the samples are obtained, on the basis of which the invention is to be developed. Taking these moral considerations into account, the Council decided to adopt a Directive on 27 November 1997.[20] However, in regard to moral safeguards the

---

12 Discussed generally hereafter.

13 Until Dec 1992.

14 Meetings were held in Oct 1994 and on 12 and 23 Jan 1995.

15 Council Common Position of 26 Feb 1995.

16 OJ C68, 20 Mar 1995. This was the first time that the European Parliament rejected a Conciliation Committee report.

17 COM (95) 661 final, 8 Oct 1996.

18 Such as Greenpeace, Friends of the Earth and religious groups.

19 See Matthias Herdegen, 'Patenting Human Genes and Other Parts of the Human Body under EC Biotechnology Directive', 2000/2001 vol. 3 *Bio-Science Law Review* 102–7 at 107.

20 *Irish Times*, 27 Nov 1997.

new proposal by the Commission was, yet again, insufficient to meet the concerns of the Parliament, and amendments were introduced. Agreement was eventually reached, and, having regard to the proposal from the Commission,[21] the Opinion of the Social and Economic Committee,[22] and acting in accordance with the procedure laid down in Article 189b of the Treaty of Rome,[23] Directive No. 98/44/ EC was adopted by the European Parliament and Council of the EU on 6 July 1998. Its long title is: Directive of the European Parliament and of the Council on the Legal Protection of Biotechnological Inventions.[24] However, behind the façade of formal agreement there remain deep divisions of opinion, the weakening of earlier opposition to the Directive reflecting not so much a revision of principle as a practical accommodation with the commercial importance of patentable biotechnology.[25]

The European Patent Office (henceforth EPO) incorporated the Biotechnology Directive into the EPC 2000 (Rule 26 *et seq.*). (In this regard, see hereunder for more details.) The purpose was to provide a supplementary means of interpretation of the EPC.

It is intended that the new Directive will enable all interested parties, including the EPO and national patent-granting authorities, to interpret in the same way those provisions of patent law that apply to biotechnological inventions. Hence, it is much easier to distinguish between what is patentable and what is not. In this sense, harmonization of patent law is facilitated.

However, the Biotechnology Directive is binding in national law only. Therefore, the obligations of EU States concerning, in particular, the moral problems of patenting the processes and products of biotechnology are likely to continue. And, because the Directive continues the shortcomings of Article 53 EPC, what it adds in terms of protection deserves careful consideration. The policy underlying the Directive is suspect, and the likely impact of the Directive on patenting biotechnological inventions in the future, uncertain. In terms of protection, harmonization under the Directive can be achieved only when Member States either reach a consensus as to the meaning of *ordre public* and morality within the context of patentability, or reject it altogether.

---

21   OJ C 296, 8 Oct 1996 at 4, and OJ C 311, 11 Oct 1997 at 12.

22   OJ C 295, 7 Oct 1996 at 11.

23   Opinion of the European Parliament of 16 July 1997, OJ C 286, 22 Sept 1997 at 87. Council Common Position of 26 Feb 1995, and Decision of the European Parliament, OJ EC 30 July 1998.

24   Directive came into force on 30 July 1998, date of publication in the OJ EC.

25   See R Brownsword, W R Cornish and M Llewelyn, Editors comment: 'Human Genetics and the Law: Regulating a Revolution', 1998 vol. 61 *Modern Law Review* 593–7 at 594.

States' Obligations and Moral Criteria of Patentability

Both the European Patent Convention 1973 and the Directive on the Legal Protection of Biotechnological Inventions 1998 contain a morality clause precluding from patentability inventions in certain circumstances. As already noted, excluded from patentability under the EPC are inventions whose *exploitation* would be contrary to *ordre public* or morality.

The provision in the Directive equivalent to Article 53(a) is outlined in Article 6, which reads partly as follows: 'Inventions shall be considered unpatentable where their commercial exploitation would be contrary to *ordre public* or morality; however, exploitation shall not be deemed to be so contrary merely because it is prohibited by law or regulation.' The problem is the Directive is binding in national law only, suggesting that patent-granting offices are likely to have a national approach to interpreting the provision. In addition, because the EPO is also the intended arbiter of what offends against the provision, further problems arise. Because the EPO is not an EU body, it is not legally bound to follow the EU Directive. In order to avoid the possibility of having two different patent laws, namely, EPC and Directive law, coexisting in European states that are EPC and EU members, the EPC will have to be adapted to the provisions of the Directive.[26] However, even if such an event occurs, it may not result in harmonization of patent law in the field of biotechnology across the states involved. The reason is, in national and EPC law, the concepts of *ordre public* and morality are likely to be interpreted in different ways, resulting in divergence of practice among Member States.

Language of the Directive

While the text of the EPC is silent as to what constitutes *ordre public* and morality, the Directive offers guidance in relation thereto and lists what is not patentable by virtue of *ordre public* or morality. Article 6(2) reads:

On the basis of paragraph 1,[27] the following, *in particular*,[28] shall be considered unpatentable:

(a) processes for cloning human beings;

26　The EPO fully intends to do this in all instances where it is necessary, Dr Siobhan Yeats, Examiner at the EPO, speaking at *The Conference of Law and Genetics*, 22 Jan 1999, Signet Library, Edinburgh.

27　That is, *ordre public* or morality.

28　Discussed hereunder.

(b) processes for modifying the germ line genetic identity of human beings;

(c) uses of human embryos for industrial and commercial purposes;

(d) processes for modifying the genetic identity of animals which are likely to cause them suffering without any substantial benefit to man or animal, and also animals resulting from such processes.

In three respects the language of the Directive is problematic such that biotechnological inventions are unlikely to receive the protection they deserve. First, by adopting the phrase 'in particular' the Directive suggests that the list of unpatentable subject-matter is illustrative rather than exhaustive. The Directive permits patent-granting offices to exclude from patentability a very wide range of inventions on the basis of morality. In respect of *public ordre* and morality concerns, the Directive expands the jurisdiction of national patent-granting offices beyond that envisaged for the EPO by virtue of Article 53(a) EPC. The addition of Article 6(2) in the Directive is a fresh provision intended to provide guidance on the meaning of *ordre public* or morality. However, in comparison to EPO jurisprudence already established by virtue of Article 53(a) EPC, Article 6(2) of the Directive adds new doubt in relation to what is patentable. Second, the Directive makes no direct reference to plant-related inventions constituting non-patentable subject-matter. However, under the Directive plant-related inventions are excluded from patentability on the basis that arguments against patenting animal-related (outlined in Article 6(d)) and plant-related inventions run in parallel. Given the difficulty the EPO has experienced in interpreting Article 53(b) EPC in relation to plant-related inventions,[29] an opportunity was missed in the Directive to clarify the law in this area. Third, the Directive makes no direct reference to the environment.[30] Given public concern over environmental issues, it is regrettable the Directive did not include environment as a factor by which *ordre public* or morality could be evaluated. However, omitting environmental concerns could be interpreted as meaning that such issues are not moral-related. If the latter, the Directive runs counter to principles enunciated by the Technical Board in *Onco-Mouse*.[31]

Policy Underlying the Directive

The Directive provides some guidance on what is, and is not, considered to be contrary to morality. In this regard, as already mentioned, the Directive has been incorporated into the EPC 2000. In the context of morality, Article 5 and Article 6

---

29   See generally Ch. 4 heretofore.

30   While the operative part of the Directive makes no direct reference to the environment, Recital 10 does.

31   [1991] EPOR 525. See generally Ch. 4 heretofore.

of the Directive are of particular relevance and have been incorporated into EPC 2000 as Rules 28 and 29.[32]

Rule 28 deals with exceptions to patentability and reads as follows:

1. processes for cloning human beings;
2. processes for modifying the germ line genetic identity of human beings;
3. uses of human embryos for industrial or commercial purposes;
4. processes for modifying the genetic identity of animals which are likely to cause them suffering without any substantial medical benefit to man or animal, and also animals resulting from such processes.

Rule 29 deals with the human body and its elements and reads as follows:

(1) The human body, at the various stages of its formation and development, and the simple discovery of one of its elements, including the sequence or partial sequence of a gene, cannot constitute patentable inventions.
(2) An element isolated from the human body or otherwise produced by means of a technical process, including the sequence or partial sequence of a gene, may constitute a patentable invention, even if the structure of that element is identical to that of a natural element.
(3) The industrial application of a sequence or a partial sequence of a gene must be disclosed in the patent application.

The Directive was drafted with the underlying idea that patentability requirements should be kept in conformity with the original EPC 1973,[33] and in this regard exclusions from patentability under the EPC, and now the EPC 2000, are continued in the Directive. As already noted, Article 6 of the Directive permits exclusion on a basis similar to that outlined in Article 53(a) EPC; and Article 4 permits subject-matter exclusion similar to that outlined in Article 53(b) EPC. On this basis, will EPO difficulties in the interpretation of Article 53 exclusion provisions continue under the Directive? Some of the difficulties arising can be seen in the approach of the EPO.

Rule 28 and Rule 29 have been subjected to discussion. In the context of stem cell patent applications, recently questions have been referred to the Enlarged Board of Appeal for clarification on the interpretation of Rule 28(c). The *WARF*[34] application relates to cell cultures comprising primate embryonic stem cells. The only starting material disclosed in the application was pre-implantation embryos. The Examining Division was of the opinion that Rule 28(c) should be interpreted broadly and was not only relevant to the claimed subject-matter but to the whole

---

32   Previously these were Rules 23(d) and 23(e) under the EPC 1973.

33   See the Explanatory Memorandum to the Directive COM (95) 661.

34   European Patent Application 1995 filed by Wisconsin Alumni Research Foundation; Technical Board of Appeal referral: T 1374/04.

disclosure. Therefore, the application was refused on the basis that the invention involved the direct and unavoidable use of human embryos. On appeal, the Technical Board of Appeal referred the matter to the enlarged Board of Appeal. Among the questions asked was: does Rule 28(c) EPC apply to an application filed before the entry into force of the Rule? On a second question, does the same portion of the Rule forbid the patenting of claims directed to products which (as described in the application) at the filing date could be prepared exclusively by a method which necessarily involved the destruction of human embryos?

The Decision by the EBA was issued on 4 December 2008. It answered the two questions in the affirmative. Thus, the claim in *WARF*'s application was held by the EBA to violate Rule 28(c) EPC.[35]

The decision was based closely on the specific facts of the case, and so, the door has not been closed on the grant by the EPO of European patents relating to human embryonic stem cells. In this respect, it is likely that the morality provision of the Directive will be interpreted in keeping with EPO jurisprudence. However, a close examination of the Directive reveals that it is concerned with moral principles on a very broad basis.[36]

The Directive offers national courts and patent offices a general guide to interpreting *ordre public* and morality.[37] Unless these concepts are interpreted in a uniform manner across all Member States, and in accordance with EPO jurisprudence, the result is likely to be a divergence of patent law and practice. As far as national law is concerned, both doctrine and courts suggest that *ordre public* is a body of positive law. The question is: what kind of legal provisions qualify under this body of law? Positive law in this context suggests criminal law, constitutional law or other special laws protecting human life and dignity, as well as other basic values in society.

As far as national patent law is concerned, there is no uniform concept of either *ordre public* or morality, and under the EPC these concepts are, allegedly, a matter for European institutions, namely, the EPO. However, the Directive attributes the joint concepts of *ordre public* and morality with a dimension wider than what states intended originally under Strasbourg and, later, the EPC. If correct, the Directive, designed to harmonize, may result in greater uncertainty for breeders and inventors in this area.

---

35   The announcement of the EPO website states: 'The EBA stressed that its decision does not concern the general question of human stem cell patentability'.

36   It is argued that the Directive has created a new and important source of law which is distinct from both national and EPO case law, resulting in clarity in the legislative framework and a legal predictability and certainty (in knowing what can be patented and what the parameters of protection are). See Tony Howard, 'The Legal Framework Surrounding Patents for Living Materials', in Johanna Gibson (ed.), *Patenting Lives: Life Patents, Culture and Development* (2008) Ashgate Publishing, UK.

37   Recital 38.

Concerns under the Directive

As already noted, Article 6[38] of the Directive recites, in effect, Article 53(a) EPC and prohibits patenting of inventions whose commercial exploitation would be contrary to *ordre public* or morality. The Directive, in Recital 38, recognizes that these concepts concern 'ethical or moral' principles. However, difficulty lies in determining what such principles embrace.[39] Setting ethical boundaries of acceptable technology is the job of governments. Support for this comes from the view of Professor Gerald Dworkin, who, in his 1994 evidence to the House of Lords Select Committee on the European Communities,[40] said:

> Few would deny that there are major ethical issues relating to developments in biotechnology and genetic engineering; that there is a need to ensure that such ethical issues are properly addressed; that there should be adequate controls and monitoring of undesirable or questionable developments ... [t]he real question, though, is whether such control should be exercised in any significant way through the patent system. A rational answer must be 'no'.

In addition, examiners at the EPO are not sufficiently trained to tackle ethical issues. Support for this is found in the view of the House of Lords Select Committee on the European Community, where it said in 1994:[41] 'We recognise, however, that patent examiners should not be faced with complex ethical issues requiring the balancing of conflicting rights or interests where the general public is itself divided.' An additional difficulty arises by virtue of the fact that the Directive is an attempt to mix Community-made law with legal obligations under an international treaty wider than the Community.[42] Support for this is found in both Recital 43, which recognizes the European Convention on Human Rights (henceforth ECHR) as forming part of the general principles of Community law; and Recital 16, which recognizes that patent law must be applied so as to safeguard the *dignity* and *integrity* of the person. Does this mean that in assessing patent applications moral criteria should be based on the ECHR?[43] Can the ECHR be distinguished from other types of conventions, in that it was drafted to accord protected human interests a special status? A positive answer suggests respect for human rights in interpreting all subsequent conventions. On this basis, even if there were no morality provision in the Directive, resolution of morally problematic concerns would be based on the provisions of the ECHR. A 'rights' based approach focuses

38   Art. 6(1).
39   See Derek Harms, 'Drafting Claims around Morality', 1996 vol. 7 *European Intellectual Property Review* 424–5.
40   See HL Paper 28, HMSO, 1 Mar 1994.
41   Ibid. at para. 46.
42   The EPC.
43   See Richard Ford, 'The Morality of Biotech Patents', *supra* at 317.

on the interests of the individual and is designed to protect the individual from claims that harming them is justified in the interests of society or of others.[44] A balance is required.

How, therefore, is the Directive to be interpreted? If the Directive is interpreted in accordance with moral criteria based on the ECHR, the likely result is a conflict with existing jurisprudence, not only between national patent-granting offices, but also within the EPO.[45] EPO jurisprudence suggests that the morality provision be interpreted in a restrictive manner. However, if the ECHR is binding on patent-granting authorities, then failure to apply a vigorous moral assessment is illegal. Although ECHR rights are usually carefully balanced between 'private' and 'public', the application of ECHR standards may be inappropriate for the following reasons. First, in respect of patentability, existing EPO jurisprudence might be abandoned with the result that patentees would no longer be certain when their actions were safe and lawful. Second, since the ECHR is based on respect for human rights only, failure to consider other concerns relevant in a commercial environment, such as the economic and social impact of biotechnology, may be itself immoral because moral decision-making involves a continuous accommodation of conflicting values.[46]

Article 6 of the Directive is reinforced by Recital 38. This states that Article 6 contains a list of inventions excluded from patentability so as to provide national courts and patent offices with a general guide to interpreting the reference to *ordre public* and morality. However, Recital 38 proceeds to state that processes the use of which offend against human *dignity*,[47] such as processes to produce chimeras from germ cells or totipotent cells of humans and animals, are *obviously* also excluded from patentability. Recital 38 raises serious issues that need to be resolved if the Directive is not to be expanded to include wider concerns of a moral nature associated with biotechnology;[48] the question is: should it?

Article 6(2)(d) lists what in particular is unpatentable on the basis of *ordre public* or morality, namely, 'processes for modifying the genetic identity of animals which are likely to cause them suffering without any substantial medical benefit to man or animal, and also animals resulting from such processes'. In this regard, three comments can be made.

First, there is no reference to plants in the provision. This suggests genetic modification of plants is acceptable and will, presumably, be determined on normal patentability criteria. The implication here is that there exists a higher standard of morality in relation to animal-related inventions than for plant-related inventions. Is the Directive drawing a distinction between animal and plant life?

---

44   See Jonathan Herring, *Medical Law and Ethics* 2nd edn (2008) Oxford University Press.

45   See Richard Ford, 'The Morality of Biotech Patents', *supra* at 316.

46   Ibid. at 318.

47   Discussed more fully hereunder.

48   See *Relaxin* [1995] EPOR 541, discussed hereafter.

If yes, on what principle? On the one hand, EPO jurisprudence suggests that plant and animal varieties should be equated so that the principle of equal treatment applies.[49] However, on the other hand, the imposition of a higher standard of morality for animal-related inventions will result in conflicting EPO decisions. In this sense the Directive creates uncertainty for breeders as to the state of the law.

Second, embedded in Article 6(2)(d) is the principle of proportionality. However, the provision offers no guidance in relation to interpretation. How are the various elements incorporated in a proportionality test to be evaluated? How is the suffering of animals to be assessed? There are no principles common to the Member States on which to base 'risk assessment'. Presumably therefore, in the absence of common evaluation criteria, the balancing test outlined by the EPO in *Onco-Mouse*[50] will continue to apply. However, since the *Onco-Mouse* decision the EPO has ruled the balancing test is not appropriate in all types of situation. In *PGS*,[51] for example, the test was deemed inappropriate since the threats to the environment were not sufficiently substantiated at the time the EPO was asked to revoke the patent. This suggests that, where the proportionality test is deemed inappropriate, the balance of grant will lie in favour of the patentee.[52] In terms of assessment of animal suffering, the absence of common principles among Member States may render genetic modification of animals patentable, on the basis that it is unjust to deny a patent in a situation where the onus of proof cannot be sufficiently discharged. Under Article 6(2)(d), only processes likely to cause animal suffering are prohibited from patentability. Thus, in the absence of sufficient proof of suffering, such processes are patentable.

Third, the provision also raises the issue of what constitutes a 'substantial' benefit? The answer to this is found in Recital 45, which merely lists benefits to either man or animal in terms of research, prevention, diagnosis or therapy as constituting substantial benefit. Since invocation of the Recitals is required in the interpretation of the Directive,[53] and because the Recitals are drafted in broad terms, this suggests that the meaning attributed to *ordre public* and morality be extended accordingly. By comparison with Article 53 EPC, the Directive, by virtue of Article 6 and accompanying Recitals, enjoys an extended moral control.

Likewise, Article 5 of the Directive encompasses moral considerations because it essentially prohibits patenting of the human body. Article 5(1) reads as follows: 'The human body, at the various stages of its formation and development, and the simple discovery of one of its elements, including the sequence or partial sequence of a gene, cannot constitute patentable inventions.' The provision is intended to preclude the grant of patents for speculative applications presenting experimental

---

49   See *PGS* [1995] EPOR 357 and Ch. 4 heretofore.
50   [1991] EPOR 525.
51   [1995] EPOR 357.
52   As the onus of proof lies with the defendant.
53   This is true for all Directives.

data based entirely on DNA sequencing.[54] DNA sequences encoding proteins of unknown function are, therefore, unpatentable. Article 5(1) accords with Recital 23, namely, that a mere DNA sequence without an indication of function does not contain any technical information and, on this basis, is not a patentable invention.[55] Recital 23 is likely to be problematic for patent-granting authorities; can it be said (with certainty) that DNA sequences per se do not contain technical information?

Recital 24 is also likely to be problematic for patent-granting authorities in the following manner. Article 5(1) accords with the requirement set out in Article 5(3), namely, that industrial application of a gene sequence or partial sequence must be disclosed in the patent application. Industrial applicability has long been accepted as a criterion of patentability. However, under Recital 24, in order to comply with the industrial application criterion, it is necessary in cases where a gene sequence or partial sequence is used to produce a protein or part of a protein, to specify which protein or part protein is produced or what function it performs. This suggests the test of industrial application is of a higher standard in cases involving gene sequences than for other inventions.

Article 5 states that a *simple* discovery of an element of the human body does not constitute a patentable invention. Article 5(1) is silent on the reason why the *simple* discovery of elements of the human body is unpatentable. Prima facie, the exclusion is justified on traditional patentability criteria, namely, discoveries per se are not patentable. Can the exclusion be justified, in addition, on moral grounds? Article 5(2) tells us[56] that elements *isolated* from the human body may constitute a patentable invention. This suggests the process of isolation is sufficient to transform a non-patentable discovery into a patentable invention. However, it can forcibly be argued that a gene is no less a gene merely because it is isolated from the human body.

Article 5(2) tells us that elements isolated from the human body, or otherwise produced by means of a technical process, may constitute a patentable invention. The provision has two limbs. First, the process of isolating and transferring a gene into a host cell provides a means for industrial application. This is where the invention lies. Identification of a gene sequence and the protein for which it encodes is merely a discovery as long as the gene remains in the human body since at this stage man cannot

---

54   Dr Siobhan Yeats, Examiner at the EPO, 'The Impact of the Biotechnology Directive on Patent Law and Practice in the Field of Genetics', *supra*.

55   There is a proposal to introduce a *genetic sequence right* which would not come under the control of any single person or organization. It recognizes that irrespective of whether a genetic sequence is an 'invention' or not, elucidation and identification of its function is important work and should be encouraged, thus enabling the holder to receive a GSR fee in return for disclosure, rather than conferring exclusive rights as in the patent system per se. See Luigi Palombi, 'The Genetic Sequence Right: A *Sui Generis* Alternative to the Patenting of Biological Materials', in Johanna Gibson (ed.), *Patenting Lives: Life Patents, Culture and Development* (2008) Ashgate Publishing, UK.

56   Among other things.

make the protein. But isolating the gene and transferring it to a new environment where the protein can be produced in a factory-like manner is inventive.[57] This portion of Article 5(2) makes sense. However, reference in the provision to elements produced by means of a technical process is unnecessary because in genetic engineering cDNA[58] is actually produced rather than naturally occurring DNA. cDNA is always produced as a result of a technical process and in this sense it is no different from any other manufactured product as regards patentability.

Second, Article 5(2) proceeds to tell us that isolated elements, or elements produced from a technical process, do not fall foul of the novelty requirement merely because the structure of the element is identical to that of a natural element. The novelty requirement in respect of elements isolated from the human body is not 'absolute' in the strict sense; the test is not whether the substance exists, but whether it was known to exist in terms of structure.[59] However, this test is wholly inappropriate for elements produced as a result of a technical process. In such a situation the product, namely, cDNA, does not occur naturally. Article 5(2) of the Directive, by equating naturally occurring elements with elements produced as a result of a technical process, suggests a profound misunderstanding of the nature of biotechnological inventions. In this respect, the Directive is misleading and likely to result in confusion in respect of genetic engineering inventions.

Already Recital 24 is proving to be problematic for courts in the UK. In a recent Patents Court decision, *Eli Lilly and Co v Human Genome Sciences, INC*,[60] Kitchin J held a patent invalid for, *inter alia*,[61] lack of industrial application.

The claimant applied to revoke European Patent (UK) 0,939,804 held by the defendant/patentee. The patent disclosed the nucleotide and amino acid sequence of a novel member of the TNF ligand superfamily which it called Neutrokine-alpha. The application for patent was filed on 25 October 1996 and it was granted on 17 August 2005. Kitchin J outlined the applicable legal framework. He said for such inventions the relevant provisions of the EPC[62] are to be applied and interpreted in accordance with Chapter V of the Implementing Regulations[63] and the Directive is to be used as a supplementary means of interpretation. He said: 'in a nutshell, the industrial application of a gene must be disclosed in the application. If it encodes a protein then the protein or its functions must be specified'.[64]

---

57   See Ulrich Schatz 'Patents and Morality', in Sigrid Sterckx (ed.), *Biotechnology, Patents and Morality* (1997) 159–70 at 168.

58   CDNA does not contain 'introns' which are non-coding sequences. cDNA is, therefore, more pure than naturally occurring DNA.

59   See Stephen Crespi, 'Biotechnology Patenting: The Wicked Animal Must Defend Itself', 1995 vol. 9 *European Intellectual Property Review* 431–41 at 432.

60   [2008] RPC 29

61   Also held invalid on grounds of insufficiency and lack of inventive step.

62   Art. 52 EPC.

63   Rules 26–34 and Rule 42(1)(f).

64   See paras 185/186 of the decision.

Kitchin J, while noting that there was very little authority from the UK jurisdiction on the scope of the industrial application requirement, nevertheless proceeded to give a good statement of the law in the UK.[65] He then outlined the approach adopted in the EPO as developed through a number of decisions.[66] Finally, he outlined the position of US law by means of the utility requirement and agreed with the conclusion of the courts there: namely, in return for a monopoly the patentee must disclose how the invention can be used (Chapter 6 heretofore). Kitchin J said that 'a patent is not a hunting licence to find a use for the claimed product. It is a reward for the successful conclusion of the search'.[67]

Kitchin J formulated nine principles which he believed emerged from the cases he referred to. Prominent among these was that the notion of industry must be construed broadly and the description must disclose a practical way of exploiting the invention. If a substance is disclosed and its function is essential for human health, then the identification of the substance having that function will immediately suggest a practical application. If, on the other hand, the function of that substance is not known, or is incompletely understood, and no other practical use is suggested for it, the requirement of industrial application is not satisfied. Kitchin J believes these principles are consistent with the Directive and the approach adopted by US courts in considering elements of the specific, substantial and credible test in respect of the requirement of utility. He says that underlying the utility provision is the same policy consideration that applies in respect of the EPO: namely, that in return for a monopoly the patentee must make full disclosure of the invention including a *practical use* to which it can be put. The conclusion is the test of industrial application now in UK courts is of a higher standard in cases involving gene sequences than for other inventions.

The Likely Impact of the Directive on EPO Jurisprudence

The EPO has granted one patent only in respect of an invention involving the isolation of human genes for medical purposes. There, objections were raised, *inter alia*, under Article 53(a) EPC. The Opposition Division dismissed the objections. An appeal is pending before the Technical Board of Appeal. The case concerned is *Howard Florey/Relaxin*.[68]

Relaxin is a hormone capable of influencing smooth muscle contraction in the body. It is used in many medical situations, including childbirth. Human sources for obtaining medicines by genetic engineering are preferred to animal sources because of their greater immunological tolerance in the human body. The invention related to a DNA fragment able to encode a specific human relaxin. The DNA sequence

---

65   See paras 186–8 of the decision.
66   See paras 189–217 of the decision.
67   See para. 224 of the decision.
68   [1995] EPOR 541.

encoding for human relaxin was taken from a pregnant woman by surgery. The DNA was integrated into a bacterial host and later used for industrial production of the relaxin. The patent was granted in April 1991 and was later opposed. The opponents[69] argued that the claimed DNA sequence offended against Article 53(a) EPC.

The opponents' specific allegations in respect of Article 53(a) EPC were that the invention:[70]

- involved the patenting of human life, which is intrinsically immoral;
- was an abuse of pregnant women and immoral in that it constituted an offence against human dignity;
- was a return to slavery and the piecemeal sale of women, infringing the human right to self-determination.

The Opposition Division rejected the objection on all grounds. The Opposition Division found the allegation that human life was patented to be unfounded. The Division pointed out that DNA was not 'life'. Rather, it was a chemical substance carrying genetic information. Patenting a single human gene had nothing to do with patenting life.[71] Since the patenting and exploitation of other human substances was not objected to by the opponents,[72] there could be no moral distinction in principle between patenting genes and these other substances. As already noted, the Division said:[73] 'Even if every gene in the human genome were cloned (and possibly patented), it would be impossible to reconstitute a human being from the sum of its genes'. The statement by the Division runs contrary to Article 6(2) of the Directive prohibiting patenting of processes for cloning human beings.

In response to the allegation that patenting relaxin was an abuse of pregnant women and an offence against human dignity, the Division justified its decision on two grounds.[74] First, women who donated tissue consented to do so within the framework of necessary gynaecological operations. Human material had been a source of useful products to man for many years and there was no evidence to suggest that such practices were unacceptable. Second, removal of human tissue was acceptable as long as there was compliance with appropriate information and consent procedures, within the framework of the Draft Bioethics Convention of the Council of Europe.[75]

---

69 Fraktion der Grunen im Europäischen Parlament and their Fraktionspräsident, Mr Paul Lannoye.

70 See *Relaxin* [1995] EPOR 549, at point 6.1 of the Reasons for the Decision.

71 Ibid. at 6.3.4.

72 The interferons and erythropoietin, for example, as well as blood and bone, which had been widely used for many years as a source of useful products.

73 See *Relaxin* [1995], *supra* at 6.3.4.

74 Ibid. at 6.3.1.

75 See the *Convention on Human Rights and Biomedicine*, Article 13, 4 April 1997, ETS No. 164. A new Additional Protocol to the Convention on Human Rights and

In response to the opponents' assertions concerning slavery and the dismemberment of women, the Division said these arguments were based on a misunderstanding of the patent system in general and of the particular patent in issue. A patent merely confers on the patentee the right to exclude others from using the invention without consent. There is no right to individual human beings under a patent. The donor was as free to live her life as she wished, and enjoyed the same right to self-determination, after the grant of the patent as she did before grant. The invention in issue related to a cloned gene capable of producing proteins in a technical manner. The only stage at which a woman was involved was at the beginning as a voluntary source of relaxin.

As already noted, Recital 43 of the Directive states that the ECHR forms part of the general principles of Community law; and Recital 16 states that patent law must be applied so as to protect the *dignity* and *integrity* of the person. How, then, should we understand human dignity; and how does it relate to human rights? In the ECHR reliance on human dignity is merely implicit.[76] Human dignity appears in various guises: sometimes as the source of human rights; at other times as itself a species of human right; and sometimes reinforcing, at other times limiting, rights of individual autonomy and self-determination.[77] What constitutes human dignity is unclear and, for this reason, it should not be accorded a status that precludes patentability on the basis of an extended moral control. Support for this is found in the case of *Netherlands v Parliament and Council,*[78] a decision by the European Court of Justice (henceforth ECJ) upholding the validity of the Directive. There the ECJ took the view that the Directive framed patent law in stringent enough terms to ensure that the human body is unavailable for patenting and inalienable and to safeguard human dignity.[79]

The *Relaxin* case raised the issue of the commercialization of human genes. The commercialization of genetic research is growing. Knowledge so generated is increasingly being translated into commercial products. But the legal (and ethical) debate on the commercialization of genetic research is very polarized. One reason is that worldviews are in conflict as to the interplay between commercial interests are those of the public.[80] Does commercialization

---

Biomedicine was adopted on 7 May 2008.

76   See the European Convention on Human Rights, Article 3 of which concerns torture, inhuman, or degrading treatment or punishment.

77   See Deryck Beyleveld and Roger Brownsword, 'Human Dignity, Human Rights, and Human Genetics', 1998 vol. 61 *Modern Law Review* 661–80 at 665.

78   Case C-377/98, decided 9 Oct 2001.

79   Ibid. at point 77.

80   See Andrea Boggio, 'Public Domain Sharing, Patents, and Fees Resulting from Research Involving Genetic Databases' in Bernice Elger, Nikola Biller-Andorno, Alexandre Mauron and Alexander M. Capron (eds), *Ethical Issues in Governing Biobanks* (2008) Ashgate Publishing, UK.

of human genes offend against human dignity? Does dignity militate against a market in human genes? While commerce in human genes might encourage some undignified trading, this is not the same as offending against human dignity. The *Rexalin* case demonstrates objections that the patenting of human gene sequences is immoral as violating human dignity are unfounded. Could the opponents have argued that a patent on such a product compromised anyone's dignity? The likely answer is no, because the researchers did acknowledge other participants and contributors suggesting that no person's dignity was offended. Nor can it be argued that the pregnant women compromised their own dignity. To this extent at least, the Opposition Division was surely correct in dismissing the opposition.[81] However, by permitting the patent the Opposition Division condoned commercialization. This raises the question of how such transactions should be monitored in those many contexts where there is a significant difference in bargaining strength between the parties. The answer, it seems, is provided in Recital 26, which outlines the requirement of informed consent. However, it has been suggested elsewhere the principle of informed consent is no guarantor of autonomy, and, in addition, because it is not included in the Implementing Regulations 2000 UK, its legal status and effect on different Member States are uncertain.[82]

Is *Relaxin* likely to be decided differently by the Technical Board of Appeal under the Directive?

As previously noted, the EPO decided to implement the provisions of the Directive for the purposes of providing a supplementary means of interpretation of the EPC.[83] Rule 23(e), which has been inserted into Part 11 of the Implementing Regulations of the EPC, mirrors exactly the text of Article 5 of the Directive. It should be noted that the decision to adopt these new Implementing Regulations was taken notwithstanding the controversial comments of the Technical Board in *Novartis*,[84] dismissing arguments that it should refer to the Directive for the purpose of determining the scope of exclusions contained within the EPC. Given the recent ECJ decision on the validity of the Directive, it is unlikely EPO jurisprudence will be abandoned.

---

81   Deryck Beyleveld and Roger Brownsword, 'Human Dignity, Human Rights, and Human Genetics' *supra* at 675.

82   See Graeme Laurie, '(Intellectual) Property? Let's Think About Staking a Claim to Our Own Genetic Samples', HUGO Satellite Conference, Edinburgh University, Apr 2001.

83   Decision of the Administrative Council 16 June 1999 to amend the Implementing Rules of the EPC; in particular, Rule 23(b)(1) states that the Directive shall be used as a supplementary means of interpretation.

84   See *Novartis/Transgenic Plant* [1999] EPOR 123.

Additional EU Regulatory Framework on Modern Biotechnology

The EU had already established a framework offering guidance to deal with concerns underpinning biotechnology such that the extension of moral considerations outlined in the Directive were unnecessary.

In recent years, a number of proposals to place genetically modified products on the market resulted in an increased level of public awareness in relation to genetic modification. In this regard, the original Deliberate Release Directive 90/220/EEC established criteria in respect of human health and the environment to be complied with before marketing could occur; hence its importance. Concerns about environmental and human health risks exist because:[85]

- there is general unfamiliarity among the public with approval and control procedures;
- the technology involved is complex and changing rapidly;
- adequate information is not being provided to allow the public to choose between modified, and conventional, products;
- use of antibiotic-resistant marker genes may impact on the use of antibiotics in both human, and veterinary, medicine;
- making crops resistant to herbicides may increase rather than reduce the use of herbicides, with a consequent build-up of chemicals in soil and water.

The Community's *regulatory* framework on modern biotechnology was designed for the dual purpose of ensuring adequate protection of human health and the environment, and creating an internal market in biotechnological products. It is founded on six principles:[86]

- *necessity*: legislation to be brought forward only if it is shown to be necessary in the light of specific biotechnological applications.
- *efficient interaction*: biotechnologically derived products will be subject to only one evaluation criteria; product evaluation to take place in accordance with three established criteria, namely, safety, quality and efficacy.
- *adaptation to progress*: the regulatory framework will be kept up to date with scientific and technical progress.
- *standards*: the development and existence of standards may be used to complement legislation.

---

85   See the Department of the Environment and Local Government of Ireland consultation paper *Genetically Modified Organisms and the Environment* (Autumn 1998).

86   See the European Commission Communication on Biotechnology, 'Preparing the Next Stage', in the White Paper on *Growth, Competitiveness, Employment*, COM (94) 219 final, 1 June 1994.

- *international obligations*: all decisions in the field of biotechnology will be in conformity with international obligations, in particular with the provisions resulting from World Trade Organization negotiations.

By contrast, the Directive on the Legal Protection of Biotechnological Inventions 1998 has failed to address adequately one of the more important issues in the debate concerning the legal, moral and social problems associated with biotechnology, namely, risk to man and the environment. The operative part of the Directive makes no mention of the environment, nor does it lay down any principles for risk assessment; however, Recital 10 of the Directive states that regard should be had to the potential of the development of biotechnology for the environment. In particular, biotechnology should be used for the development of methods of cultivation less polluting and more economical in their use of ground.

The Biotechnology Directive, in not addressing the core problem of environmental biotechnology, suggests that existing legislation is sufficient for the adequate protection of moral concerns in this area.

Deliberate Release Directive 1990

The overall objective of the original Deliberate Release Directive[87] was to approximate the laws, regulations and administrative provisions of the Member States in relation to the deliberate release of genetically modified organisms (hereafter GMOs), and to protect human health and the environment in the context of such releases.[88] The Deliberate Release Directive in its original form establishes Community-wide procedures for obtaining consent for the deliberate release of GMOs to the environment. It also lays down a harmonized system of procedures and conditions for all deliberate releases based on prior environmental risk assessment and evaluation. (However, the Directive in its original form does not specify common principles for environmental risk assessment. As noted hereafter, this is a weakness within the framework of the original Deliberate Release Directive).[89] By contrast, the Biotechnology Directive does not lay down a system of procedures and conditions in relation to patentability criteria.

Part A of the Deliberate Release Directive requires Member States to designate a competent authority for the purposes of implementing the provisions of the Directive. Proposals to release GMOs deliberately must be submitted to the competent authority, and prior written consent must be obtained therefrom. The notifier must present a comprehensive environmental risk assessment in each case for the purposes of evaluation by the competent authority concerned. For the

---

87   Repealed and now replaced by Council Directive 2001/18/EC.

88   See the Department of the Environment and Local Government of Ireland consultation paper *Genetically Modified Organisms and the Environment, supra.*

89   Discussed hereafter in the context of an amendment to the Directive.

purposes of the Directive, environmental risk assessment is defined as 'Evaluation of the risk to human health and the environment (including plants and animals) connected with the release of GMOs or products containing GMOs'.

Part B of the Directive requires the notifier to seek consent for deliberate release into the environment for research and development purposes, or for any purpose other than placing on the market from the competent authority where release is to be effected. Therefore, Part B procedures operate only at the national level.

Part C of the Directive deals specifically with proposals to place products containing GMOs on the EU market. Consent to market under Part C is given only if consent under Part B has already been obtained.

Where new information becomes available regarding the risks posed by the product, either prior to, or after, consent has been granted, the notifier is required to inform the competent authority, and to take measures necessary to protect human health and the environment. Disclosure of information relating to notifications is subject to certain restrictions in respect of commercially sensitive data and to intellectual property rights. It is a matter for the competent authority, in consultation with the notifier, to decide which information is to be treated as confidential. However, the following cannot be treated as confidential:

- description of the GMO, name and address of notifier, purpose and location of release;
- methods and plans for monitoring the GMO, and for emergency response;
- evaluation of foreseeable effects, especially any pathogenic or ecologically disruptive effects.

The EU White Paper on *Growth, Competitiveness and Employment* 1994[90] stated that horizontal approach to regulatory control is unfavourably perceived by scientists, and the biotechnology industry, as having a restraining effect on competitiveness. The Paper recommended that: 'The Community should be open to review its regulatory framework with a view to ensuring that advances in scientific knowledge are constantly taken into account and that regulatory oversight is based on potential risks'. On this basis, the Commission undertook to review the Deliberate Release Directive with a view to:

- extending its flexibility, so that its scope and the procedures to be followed are always appropriate to the risks involved, and are easily adaptable;
- strengthening more uniform decision-taking between Member States in the case of research and development releases;
- introducing further opportunities for notifiers so that they can benefit more from the existence of a uniform Community system;
- facilitating the link between the Deliberate Release Directive and product legislation.

---

90   COM (94) 219 final.

The report concluded that implementation of the Directive revealed a number of problem areas, including:

- insufficient clarification concerning the objectives for risk assessment, which has hindered full harmonization between Member States at the research and development stages, and which has led to disagreements between Member States, at the stage of placing on the market of products.
- absence of a link between administrative procedures and identified risk;
- a weak link between parts B and C of the Directive, which means that research and development releases under Part B do not always provide the relevant data for the environmental assessment necessary for the placing on the market under Part C.

In view of these conclusions, the Commission undertook to adopt a proposal for an amendment to the Deliberate Release Directive. The main elements of the amendment involve:

- general provisions to clarify the scope of the Directive, by providing specifically for both direct and indirect environmental issues to be addressed in the environmental risk assessment;
- provision for the Commission to consult any committee it has or may establish, for the purpose of obtaining advice on ethical considerations;
- common principles for environmental risk assessment, to apply both to research and marketing releases.

The original Deliberate Release Directive offered guidance for the Biotechnology Directive in two respects. First, under the Biotechnology Directive, moral issues could have been dealt with as they are in the Deliberate Release Directive, that is, at the pre-application stage. This would enable patent-granting authorities to grant patents on clear patentability criteria. Such an approach accords with Article 6(1) of the Directive, which, as already noted, in not using the term 'publication', suggests removal of discretion from patent-granting offices to evaluate the morality of creating inventions. In the interests of harmonization, a similar approach could have been adopted by the EPO. Second, the introduction of common principles of risk assessment would greatly reduce the possibility of conflict among Member States as to the meaning and application of moral concerns. In this manner, harmonization of patent law and practice would be enhanced.

On 12 March 2001 the second Deliberate Release Directive was adopted.[91] It clarified the scope of the original Directive and of the definitions therein. The new

---

91   Directive 2001/18/EC of the European Parliament and of the Council on the deliberate release into the environment of genetically modified organisms.

Directive takes into account the so-called 'precautionary principle',[92] which, it is suggested, if applied to itself, would likely never have been adopted, for it makes little sense. More importantly, the new Directive takes into account principles for environmental risk assessment which are both real and effective.[93] Article 1 outlines the objective of the Directive and reads as follows:

> In accordance with the precautionary principle, the objective of this Directive is to approximate the laws, regulations and administrative provisions of the Member States and to protect human health and the environment when:
>
> – carrying out the deliberate release into the environment of genetically modified organisms for any other purposes than placing on the market within the Community,
>
> – placing on the market genetically modified organisms as or in products within the Community.

Article 2 (8) tells us that 'environmental risk assessment' means the evaluation to risk of human health and the environment, whether direct or indirect, immediate or delayed, which the deliberate release or the placing on the market of GMOs may pose and carried out in accordance with Annex ii.

Annex ii lays down the principles for the environmental risk assessment. It describes in general terms the objective to be achieved, the elements to be considered and the general principles and methodology to be followed to perform the environmental risk assessment.

The new framework, underpinning as it does moral concerns, suggests that the role of the morality provisions in the Biotechnology Directive is now questionable.

Conclusions

The Directive on the Legal Protection of Biotechnological Inventions 1998 aims to harmonize the criteria for the patentability of genetic material across EU Member States and to facilitate a uniform application of the immorality exclusion in all Member States so as to encourage investment in biotechnology. However, since the obligations of States of the EU regarding the moral criteria of patentability

---

92   The precautionary principle is a moral and political principle which states that if an action or policy might cause severe or irreversible harm to the public or to the environment, in the absence of a scientific consensus that harm would not ensue, the burden of proof falls on those who would advocate taking the action.

93   Recital 8 tells that the precautionary principle has been taken into account in the drafting of this Directive and must be taken into account when implementing it.

are conflicting between the EPC and the Directive, this is unlikely to occur. The Directive perpetuates the shortcomings of Article 53 EPC, suggesting that it, too, is not an appropriate mechanism for determining exceptions to patentability. If the Directive was drafted to accord more fully with existing Community legislation uniform protection might be enhanced (Chapter 8).

The Directive was drafted with the underlying idea that patentability requirements should be kept in conformity with the EPC and its current interpretation at the EPO. However, as long as patentability criteria include moral concerns there is little prospect of harmonization in this area. Harmonization can only be achieved when Member States either reach a consensus as to the meaning of morality within the context of patentability or reject it altogether.

The Recitals of the Directive on the Legal Protection of Biotechnological Inventions 1998 expand the ambit of morality beyond that envisaged under Article 53 EPC. This could result in inventions pertaining to a wide range of technology, not previously excluded from patentability, being precluded under the Directive. Whether or not this is a good thing is questionable. On the one hand, if intellectual property rights are not to be placed above the right of all human beings to live a full and productive life, laws and policies regarding exclusionary rights are not only necessary but require re-evaluation. However, in the light of the European Court of Justice decision[94] upholding the validity of the Directive (Chapter 9) there is unlikely to be a legislative return to issues, including moral ones, for some time. This suggests there is no (immediate) prospect of certainty in the law for biotechnological inventions and biotechnologists' problems seem set to continue.

---

94   Judgment in Case C-377/98, *Netherlands v Parliament and Council*, was given on 9 Oct 2001.

Chapter 8
Plant Variety Protection: UPOV, Community, Directive

Introduction

The Strasbourg Convention 1963 established a patent system essentially universal in character. However, as already noted,[1] some exceptions to patentability, including plant varieties, were permitted under Article 2 thereof. The rationale underlying the exceptions is that plant varieties are best protected by specific national legislation, or, alternatively, by virtue of the International Union for the Protection of New Varieties of Plants (UPOV) Convention 1961. There was also a desire among states not to interfere with agricultural or horticultural cross-breeding processes, and the limitation on patenting was extended to essentially biological processes for the production of plants and animals. This reflected states' practice and was the intention of the drafters at Strasbourg. The result of the Strasbourg Convention was that rights became available to inventors on a set of uniform (if not clearly defined) principles. The logic of Article 2 Strasbourg was extended in Article 53 of the European Patent Convention (henceforth EPC) to include animal varieties. As already noted,[2] difficulties with the provision include that Article 53(b) constitutes an exception to the general provision of Article 52 EPC outlining patentability criteria. Since EPO jurisprudence is unclear in respect of what is unprotectable by virtue of Article 53 EPC,[3] and because under the revised UPOV Convention 1991 certain plant varieties are not eligible for protection, some life forms are without any protection. This situation needs to be remedied.

The Directive on the Legal Protection of Biotechnological Inventions 1998 (henceforth the Directive) was drafted with the underlying idea that patentability requirements should be kept in conformity with the EPC.[4] As already noted,[5] Article 4 of the Directive permits subject-matter exclusion on a basis similar to that outlined in Article 53(b) EPC. Therefore, the uncertainty in interpretation of Article 53(b) is likely to continue under the Directive and lack of protection for certain life forms will continue in the EU in the future.

---

1   See generally Ch. 2 heretofore.
2   See generally Ch. 4 heretofore.
3   See generally Ch. 4 heretofore.
4   See generally Ch. 7 heretofore and also the Explanatory Memorandum to the Directive (COM) 95 661.
5   See generally Ch 7. heretofore.

The reasons why plant variety rights systems form a part of the critique of the Directive include:

- The essence of plant breeding is the creation of genetic variation in a plant species, and the selection from within that variation of plants with desirable traits, which can be inherited in a stable fashion. In this sense, plant variety rights systems form an integral part of biotechnology.
- Arguably, biotechnology, including plant biotechnology, should be protected by one legal regime only.[6] This suggests the Directive, in perpetuating a dual system of protection, is inadequate.
- The distinction between plants and animals is not always clear in that some organisms do not possess all the features that are characteristic of the kingdom to which they have been allocated.[7] The extent to which plant and animal protection systems overlap requires examination.
- The Article 2 exceptions of the Strasbourg Convention reflecting states' practice and containing a morality clause are continued in Article 53 EPC.
- The Biotechnology Directive includes a morality clause and permits subject-matter exclusion from patentability similar to Article 53 EPC.
- By contrast with the patent system, under UPOV 1991 breeders' rights are available on a set of clearly defined principles.
- The introduction in the Community plant variety rights scheme of a morality clause restricts the exercise of rights in certain circumstances. A Community plant variety right, once granted, does not permit the rights holder to *use* the right for any purpose whatsoever. Member States may restrict the exercise of the right by national legislation. By contrast, the Directive contains an absolute prohibition on patenting plant varieties; Member States retain no discretion in this regard.
- Because the judiciary in the UK is reluctant to adapt traditional patent law criteria to deal with new problems presented by biotechnological inventions, protection of variety-related inventions there is likely to occur by means of the plant variety rights system.

The issue of what is a plant and what is a plant variety is outlined briefly as follows.[8] Traditionally, living matter is divided into two 'kingdoms': the 'animal kingdom' and the 'plant kingdom'. Plants are distinguished from animals by their structure, source of nutrition and mobility. Plants are 'autothropic' in that they synthesize food for themselves from inorganic material, using sunlight as a source of energy. Considerations based on cell structure also suggest that two other 'kingdoms' exist: the *prokaryota* and the *eukaryota*. The former refers to single-celled organisms

---

6   Currently a dual system of protection operates, namely, the patent and plant variety rights regimes.

7   See *In re Arzerberger* 27 CCPA 1315 at 1318.

8   *UPOV General Information* publication no. 408(E) 1995, ISBN 92-805-0282-4.

without differentiated nuclei, while the latter to all other organisms. The need to make information available in relation to plants has resulted in a hierarchical classification whose main levels are termed in Latin: *divisio, classis, ordo, familia, genus* and *species*.

The species is the basis of the classification. Generally speaking, 'species' denotes a group of organisms sharing a large number of heritable characteristics, and can interbreed, but are genetically isolated from other organisms by 'sterility barriers' (these prevent interbreeding). Certain species, especially cultivated ones, comprise distinct types.[9] These are taken into account in nomenclature through various 'intraspecific' levels, such as *subspecies, varietas* or *forma*. Types chosen on the basis of a subdivision within the species or the taxonomic unit of the lowest known rank are referred to as a 'variety' of that species. Plants chosen on the basis of the taxonomic unit of lowest known rank are termed a 'plant variety'. The precise nature of a group of plants comprising a 'variety' depends on factors such as: the mode of propagation of the plants; their floral biology; and the plant breeding techniques used.

Different types of plant variety have been developed, depending upon the physiology of the chosen plants, and the reproductive mode of each species. Some plants reproduce in an 'asexual' manner, or vegetatively, that is, by using part of the plant to reproduce another complete plant. Other plants reproduce in a sexual manner, where pollination of the female part of the flower (the stigma) by pollen from the male part of the same flower (the anther), or pollen from another flower of the same plant, occurs. However, many species of plants are not adapted to self-fertilization, or cannot tolerate self-fertilization through successive generations, without becoming less vigorous (suffering from 'in-breeding depression', with the result that yields will be lower and resistance characteristics reduced). Plants tolerating successive generations of self-fertilization, without losing vigour, can produce varieties based upon a single plant (or a small number of plants) which will reproduce precisely. All plants of a variety of this kind will be genetically identical. Such plants are said to breed 'true to type'.

Another category of variety is based upon the controlled cross-pollination of parent lines, in which case the resulting seed will have the genetic make-up of both parents. Such varieties are known as 'hybrids', and typically will exhibit greater vigour than the parent lines. 'Hybrid vigour' results in plants with higher yields, and greater resistance characteristics, thus ensuring better quality.

It has been estimated that, by the year 2020, the Earth's population may reach eight billion, with 83 per cent living in so-called developing countries.[10] Coupled with the expected increase in population, there is the added problem that productive farmland is being used for non-farming purposes. It is suggested that there is no option except to produce more food on less land, in order to meet the needs of the growing population. The importance of ecologically sustainable advances in

---

9   Ibid.
10   Ibid.

the productivity, and profitability, of major farming systems is obvious.[11] Plant breeders are technically well equipped to make a major contribution to meeting these challenges. The essence of breeding is to develop plants genetically modified to produce higher yields of better-quality products.

In seeking to create genetic diversity for selection purposes, plant breeders encountered many natural problems. Crosses between closely related species were often not possible because of infertility problems, and the transfer of characteristics between unrelated species was impossible. Plant breeding was time-consuming and often random. Where a trait controlled by a recessive gene is required, the presence of the gene might be hidden in the plant and apparent only in its progeny. However, the discovery of the structure of DNA, and the fact that it constitutes the genetic code governing the inheritance of the features of all living organisms, promises to lessen, if not eliminate, the constraints imposed upon plant breeders. Techniques of genetic engineering enable transfer of useful features within a short time span, not only between plant species, but plant and animal species.

The benefits to be derived from the classical selection procedures of plant breeders, and from those of the new biotechnology, are cumulative.[12] The one cannot replace the other. It is in the interest of society to ensure that systems of incentive be continued (or created), to encourage investment in both the new biotechnological and classical plant breeding.

The fact that plants inherently possess the ability to reproduce or replicate presents plant breeders with further special problems. The release to growers of propagating material (in the form of a plant variety) enables them to reproduce the variety without further recourse to the breeder for additional supplies. The breeder's customers can secure supplies of propagating material and compete with him, thus profiting from his many years of research and breeding effort.[13]

A possible solution to this problem is for breeding to be conducted by public institutions using public money, with the results of breeding work being made available to the public free of charge or on a subsidized basis from the state.[14] However, state funding alone would be insufficient to cope with the needs of modern society in respect of the numbers of varieties required for adaptation to the wide range of circumstances pertaining worldwide. Accordingly, many countries

---

11   Several scientists believe that agricultural biotechnology has great economic and humanitarian potential. See Enrico Bonadio, 'Crop Breeding and Intellectual Property in the Global Village', 2007 vol. 29 *European Intellectual Property Review* 167 at 169.

12   See Judith R Curry's comparative study, *The Patentability of Genetically Engineered Plants and Animals in the US and Europe* (1987) 5, Intellectual Property Publishing, UK. There is now the added concept of 'terminator technology', an innovation induced in response to the problems of appropriateness and weakness in existing intellectual property rights institutions. See Kanchana Kariyawasam, 'Terminator Technology as a Technological Means of Forcing Intellectual Property Rights in Plant Gernplasm: Its Implications for World Agriculture', 2009 vol. 31 *European Intellectual Property Review* 37.

13   So-called 'free riding'.

14   *UPOV General Information, supra.*

have established systems whereby exclusive rights of exploitation are granted to the breeders of new varieties of plants in order to:

- provide breeders with an opportunity to receive a reasonable return on investment;
- provide incentive for continued (or increased) investment in the future;
- recognize the economic right of the breeder to remuneration for his efforts.

Plant varieties are developed for, and are adapted to, areas with particular agro-ecological conditions. The borderlines of such areas frequently do not correspond with national frontiers. Consequently, plant breeders normally seek protection in all states where the agro-ecological conditions are sufficiently similar. Hence, the importance of regional and international systems of protection.

Regional and International Protection for Plant Material

While protection for plant-related innovations came to the fore in Europe generally in the period about 1950, proprietary rights in plant material attracted international attention a long time prior to this:

- *1883*: The Paris Convention for the Protection of Industrial Property (PIP) recognized that the concept of industrial property should extend to 'agricultural' industries.[15]
- *1932*: The International Association for the Protection of Industrial Property (AIPPI) Congress, London, discussed the need to give plant protection.
- *1934*: The London Act of PIP expressly states that 'natural products' are to be included under the head of industrial property.
- *1952*: The Wuesthoff Report, issued in Germany, supported the idea of both patent and *sui generis* type plant protection.
- *1957*: The International Association of Plant Breeders for the Protection of Plant Varieties (ASSINEL), at a conference in Paris, selected a committee of experts to examine the situation in relation to the protection of plant varieties.

In European states, the initial inclination of plant breeders was to seek protection under existing patent systems. However, a number of technical difficulties were encountered in seeking to apply the rules of a system, designed to protect technical or mechanical invention, to plant varieties which were thought not to reproduce themselves. Nevertheless, in some European states, progress was made in affording plant protection. The Netherlands adopted the Breeders' Ordinance in 1941 and

---

15   Article 1.

granted a limited exclusive right, for breeders of agriculturally important species, to market the first generation of certified seed. Germany, too, had a system of protection for breeders based on seed certification. In addition to this, the Law on the Protection of Varieties and the Seeds of Cultivated Plants Act 1953 was enacted. The Act gave breeders the exclusive right to produce seed of their varieties for the purposes of the seed trade, and to offer for sale and market such seed.

As already noted, the United States had been the first country to introduce a special form of protection for plant material.[16] The Plant Patent Act (henceforth PPA) was enacted in 1930.[17] However, the Act is not without limitations, and not all plants fall within the statutory subject-matter for which plant patents are issued. The PPA limits protection to asexually reproduced plants. The reason why this limitation exists was the belief that such plants do not reproduce 'true to type'. The PPA also excludes coverage of tuber-propagated plants such as the potato. An additional limitation, of particular concern to plant patent holders, is that parts of plants are not protected.

Other requirements for patentability under the Act include distinctiveness, novelty, non-obviousness and disclosure. The disclosure requirement of the PPA is not as strict as that for obtaining a utility patent, which requires disclosure of the 'best mode' of performing the invention. A utility patent is a national patent rooted in the Constitution itself; it is a normal type of patent issued for a new, useful and non-obvious machine, manufacture, composition of matter or process. Because breeding procedures are not descriptively repeatable to order, the requirement under the Act is that the description be as 'complete as reasonably possible'. In order to overcome the limitations of the PPA, the Plant Variety Protection Act was enacted by Congress in 1970, to encourage the development of plants which reproduce in a sexual manner.

The International Convention for the Protection of New Varieties of Plants, adopted by Diplomatic Conference[18] in Paris on 2 December 1961, provided for the first time, recognition of the rights of plant breeders on an international basis. The International Union for the Protection of New Varieties of Plants, known as UPOV,[19] is an intergovernmental organization with headquarters in Geneva, Switzerland. The Union was established by the Convention signed in Paris in 1961. The UPOV Convention came into force in 1968 and was revised in Geneva in 1972, 1978 and 1991. The 1978 Act came into force in 1981. The 1991 Act came into force on 21 April 1998. The purpose of the UPOV Convention[20] is to ensure

---

16   See generally Ch. 6 heretofore.

17   Title 35 USC ss.161–4.

18   ASSINEL (International Association of Plant Breeders for the Protection of Plant Varieties) appointed a committee of experts and was largely responsible for the Convention.

19   The acronym UPOV is derived from the French name of the organization, namely, *Union internationale pour la protection des obtentions végétales*.

20   UPOV publication no. 437(E) (Sept 1996).

that Member States of the Union[21] acknowledge the achievements of breeders of new plant varieties, by making available to them an exclusive property right on the basis of a set of uniform and clearly defined principles. Protection of new varieties of plants was important not only for the development of agriculture in the territory of Member States but also for safeguarding the interests of breeders. However, special problems arose from the recognition and protection of the right of the creator in this field, namely, limitations that the requirements of the public interest imposed on the free exercise of such a right.[22] To accommodate requirements of the public interest, derogation from the breeder's right was necessary. This resulted in the development of a number of so-called 'privileges'.

The origin of the concept of 'farmer's privilege' lies in the 1961 Act of UPOV.[23] This permitted restriction in the exercise of protected rights only if it was in the 'public interest' to do so. It was considered to be in the public interest that the special relationship between breeders and the farming community be recognized. 'Farmer's privilege' permitted farmers to retain seed from one year to the next without having to pay an additional royalty for the saved seed. The rationale behind the concept is that farmers will continue their traditional and cultural role in taking care of land. However, it was increasingly recognized in the 1980s that the privilege was open to abuse, and it was decided that some attempt should be made to allow EU Member States to curb the practice if they so wished.

Encouraging advances in productivity and profitability of farming systems are major concerns. However, the fact that plants inherently possess the ability to reproduce or replicate presents plant breeders with special problems. While authorization of the breeder is at the heart of the revised UPOV system, in the interests of the public some exceptions had to be admitted. UPOV Member States, due to the widely differing nature of agriculture and considering the varying political situations arising in these states, are entitled under the provisions of UPOV 1991, on an optional basis, to exempt the planting of farm-saved seed from the requirement of the breeder's authorization. The 1991 Act of UPOV thus maintains a balance between the rights of the breeder and those of the farming community.

The international system of plant protection established under UPOV in 1961[24] operates worldwide, and since 1994 a Community plant variety rights scheme[25] has been operative within the Member States of the EU. Additionally, the Directive on the Legal Protection of Biotechnological Inventions 1998[26] provides

---

21   The EU per se is not a member of UPOV.

22   1961 Act of UPOV the Preamble.

23   Art. 9.

24   As revised.

25   Council Regulation (EC) no. 2100/94 of 27 July 1994 on Community Plant Variety Rights.

26   Directive of the European Parliament and the Council on the Legal Protection of Biotechnological Inventions no. 98/44/EC of 6 July 1998.

a framework whereby biological inventions receive protection within the EU. The principles established under UPOV have been incorporated into the CPVR scheme, with one important difference. The Community scheme introduced a morality clause and has the effect of restricting the *exercise* of rights in certain circumstances; by contrast, UPOV did not. Many of the protection principles underlying the CPVR scheme, in turn, form the basis of the Directive, including a morality provision operating in respect of the *grant* of rights. This suggests, in terms of morality, the Directive is more restricting than the Community scheme for plant variety rights.

This chapter outlines and compares the framework of protection provided by UPOV 1991 with CPVR 1994 in respect of living plant material; and contrasts the protection offered under the 1998 Directive.

Protection under the UPOV 1991 Act

Despite the importance attaching to new varieties of plants and the initial inclination of breeders to protect using existing structures, the patent system was regarded as an inappropriate form of protection for breeders.[27] There were two main reasons for this. First, the nature of plant material was regarded as incapable of meeting patentability criteria. Because plant material is inherently capable of replicating, it is not descriptively repeatable to order. This meant that plant material was considered not capable of meeting the disclosure requirement of the patent system. Second, it was not considered to be in the public interest to permit such an extensive monopoly over plant varieties meant for the enjoyment of all. For these reasons it was felt that a *sui generis* form of protection should be introduced for the plant breeding industry; the 1961 Act of UPOV introduced the plant variety right.[28]

Because the aims of the patent and plant variety right systems are similar, namely, protection of innovation, it might be thought that the rights overlap. However, the two systems of protection are distinct, as evidenced by looking at the administration of each system. In the UK the patent system is administered by the Patent Office,[29] while the Plant Variety Rights Office administers the rights scheme.[30] In order to obtain the plant variety right, the breeder must demonstrate that his grouping of plants is distinct, uniform and stable in essential characteristics

---

27   See Margaret Llewelyn, 'The Legal Protection of Biotechnological Inventions: An Alternative Approach', 1997 vol. 3 *European Intellectual Property Review* 115–27 at 117.

28   While national *sui generis* legislation is easier to tailor to the needs and interests of local communities, lack of international recognition means it is of limited effectiveness. See Christoph Antons, '*Sui Generis* Protection for Plant Varieties and Traditional Agricultural Knowledge', 2007 vol. 29 *European Intellectual Property Review* 480.

29   Supervised by the Department of Trade and Industry.

30   Supervised by the Ministry of Agriculture, Fisheries and Food.

following propagation. It is argued elsewhere that delinking the standards of protection from a requirement to demonstrate inventive step allows a low threshold for protection.[31]

Significant differences between the two systems meant that the type and scope of protection available under each system was clearly differentiated; a bar in both the EPC 1973 and UPOV 1961 over the acquisition of protection under the other system was imposed. However, an important distinction existed in respect of the nature of each of these bars.[32] Under Article 53(b) EPC there is an absolute prohibition on the patenting of plant varieties. On the other hand, under Article 2(1) of the 1961 Act of UPOV, Member States were afforded discretion as to whether to protect breeder's rights by either a special form of protection or a patent. However, Member States who ratified the EPC had to revise this option in the light of the absolute ban contained in Article 53(b). Therefore, for all practical purposes the plant variety right became the only method of protecting plant varieties.

In order to extend the scope of protection to innovators, and to curb abuses by farmers, revision of UPOV was felt to be necessary. This resulted in the 1991 Act. Article 2 of the 1991 text outlines the basic obligation of Contracting States, namely, to grant and protect breeders' rights as provided for in the Convention. The 1991 Act is silent on the form of the breeder's right. It may be a *sui generis* right within the context of the Convention, a patent, or any other designation as long as it has the minimum substance provided for in the Convention. The 1991 Act does not contain the bar on 'double protection'. Contracting Parties are free to protect varieties not only by means of a breeder's right, but also by other titles, particularly patents. A Contracting State wishing to provide patent protection in addition to the breeder's right can stipulate whether the applicant must choose between the two forms of protection, or whether cumulative protection is allowed. If cumulative protection of this type is obtained, resolution of any conflict between the two forms of protection is regulated by the legislation, and courts, of the state where both titles were obtained.[33]

Since the bar on double protection is abolished under the revised UPOV, it can now be argued that the type and scope of protection offered by both the patent and plant variety rights systems are similar. Therefore, in light of the changes introduced in the 1991 Act of UPOV, the ban on patenting plant varieties contained in Article 53(b) EPC is no longer necessary, cannot be justified and should be removed. It has long been recognized in EPC states that plant material per se is

---

31   See Dwijen Rangnekar, 'Is More Less? An Evolutionary Economics Critique of the Economics of Plant Breeds' Rights', in Johanna Gibson (ed.), *Patenting Lives: Life Patents, Culture and Development* (2008) Ashgate Publishing, UK.

32   See Margaret Llewelyn, 'The Legal Protection of Biotechnological Inventions', *supra* at 117.

33   See Barry Greengrass, 'The 1991 Act of the UPOV Convention', 1991 vol. 12 *European Intellectual Property Review* 466–72 at 467.

capable of protection by patent.[34] The 1991 Act merely extends the scope of the breeder's right to include protection of plant varieties by patent. In this regard, concerns that adequate safeguards to protect the interests of the public do not exist are misplaced. Additionally, by offering patent protection the revised 1991 Act of UPOV accords with international trends. As already noted, in the United States the PPA 1930 was enacted to provide patent protection for plants reproducing in an asexual manner. This Act was the first, and remains the only, law passed[35] specifically to provide patent protection for living matter.

The revised text of the 1991 Act contains a definition of the term 'variety'. Article 1(vi) provides:

> 'variety' means a plant grouping within a single botanical taxon of the lowest known rank, which grouping, irrespective of whether the conditions for the grant of a breeder's right are fully met,[36] can be
>
> •   defined by the expression of the characteristics resulting from a given genotype or combination of genotypes,
> •   distinguished from any other plant grouping by the expression of at least one of the said characteristics,
> •   considered as a unit with regard to its suitability for being propagated.

The 1991 Act makes a clear distinction between the definition of 'variety', and a variety meeting the technical criteria of Articles 7, 8 and 9[37] so as to be a protectable variety under the Convention. That the definition of 'variety' is wider than 'protectable variety' is made clear by the use of the words '[i]rrespective of whether the conditions for the grant of a breeder's right are fully met'.

Under the 1991 Act it is possible to define plant variety without reference to the grant of a plant variety right. This suggests that plant material conforming to the UPOV definition of variety, regardless of whether or not such material is eligible for protection by virtue of UPOV, is excluded from patentability by virtue of Article 53(b) EPC. In the absence of patent protection, or plant variety right, where will such variety material receive protection?

Is distinguishing between protectable and non-protectable varieties in the 1991 Act justified? The reason for distinguishing between protectable and non-protectable varieties is to give meaning to the technical criterion outlined in Article 7 UPOV, namely, distinctness. Under Article 7 of the 1991 Act, a variety,

---

34   See *Ciba-Geigy/Propagating Material* [1979/85] EPOR vol. C, *and Lubrizol/ Hybrid Plants* [1990] EPOR 173.

35   In the US.

36   The Patents Act 1977 UK amended by the Patents Regulations 2000 Schedule A 2 defines 'plant variety' but does not include the words 'irrespective of whether the conditions for the grant of a breeder's right are fully met'.

37   These relate to distinctness, uniformity and stability respectively.

to be protectable, must be distinguishable from any other variety whose existence is a matter of common knowledge at the time of the filing of the application. Under previous texts,[38] to be protectable a variety had to be clearly distinguishable by one or more *important* characteristics from any other variety. Ambiguity arose by virtue of the fact that, previously, to be protectable a variety had to be distinguishable by some feature related to merit.[39] However, the UPOV Convention, even before the 1991 Act, afforded protection to any variety clearly distinguishable from other varieties irrespective of any judgment concerning its worth.[40] Thus, while the 1991 text avoids the ambiguity of the word 'important', breeders are left without any form of protection for varieties which fail to meet the definition of protectable variety as outlined. The purpose of protection is to safeguard the interests of plant breeders and to this extent inclusion of non-protectable varieties in the 1991 text of UPOV is a gap in the Convention. It is difficult to justify the distinction in the 1991 text between protectable and non-protectable varieties. To protect adequately the interests of breeders, plant varieties excluded from protection under plant variety rights legislation should, once conforming to normal patentability criteria, be eligible for patent protection.

Scope of Protection

One of the innovative features of the 1991 Act is the extent to which the scope of protection available to breeders is increased. Arguably, breeders' rights now are akin to those of the patentee. This includes conferral of provisional protection. Article 13 of the 1991 text makes it obligatory for Member States to protect the interests of the breeder during the period between the filing of the publication of the application and subsequent grant. As a minimum the right holder is entitled to equitable remuneration in respect of acts requiring the breeder's authorization once the right has been granted. However, Contracting Parties may provide that such measures shall only take effect in relation to persons whom the breeder has notified of the filing of the application and in this manner provisional protection is limited. Nevertheless, conferral of provisional protection suggests that the plant variety right system is now more akin to the patent system than previously.[41]

One of the major innovations in the 1991 Act is that 'farmer's privilege' does not arise automatically; by contrast, in previous texts authorization of the breeder was required in respect of:[42] 'The production for the purposes of commercial marketing ... [o]f the reproductive or vegetative propagating material ... [a]nd for the offering for sale or marketing of such material'. The fact that the breeder's

---

38   In particular the 1978 Act of UPOV.

39   See Barry Greengrass, 'The 1991 Act of the UPOV Convention', *supra* at 468, who argues that this was never the case and is not true.

40   Ibid. at 468.

41   EPC Art. 67 confers provisional protection on the patentee.

42   UPOV 1961 Act Art. 5.

authorization was required for the purposes of commercial marketing only meant that production for other purposes was permitted. In effect, production for use on the farm where propagating material was produced fell outside the scope of protection. The wording implicitly created the 'farmer's privilege'.

By restricting the right to the reproductive material, the drafters of the Convention ensured that the breeder retained control over the varieties' genetic information. This permitted the breeder to maintain control over variety quality and at the same time permitted the public access to other elements constituting the variety. Arguably, the restriction reduced the effectiveness of the plant variety right.[43] Additionally, that only unauthorized sales constituted an infringing act suggested that protected varieties were available for use by breeding and other public institutes to create derived varieties. The result was that, in the light of advances in agricultural and horticultural technology, the plant variety rights system did not provide sufficient protection for biotechnological inventions of this nature. The dual protection bar, coupled with the fact that farmers were permitted to retain seed of a protected variety from one year to the next, with no additional payment to the breeder, meant that the system was regarded as not providing the level of protection felt necessary. The 1991 Act remedies that situation. Article 14 of the 1991 Act extends the scope of protection available to breeders and does not create a farmer's privilege per se.

Article 14(1)(a) provides that the following acts in respect of the propagating material of the protected variety shall require authorization of the breeder:

- production or reproduction (multiplication)
- conditioning for the purposes of propagation
- offering for sale
- selling or other marketing
- exporting
- importing.

Under the section the scope of protection is extended to all production or reproduction, without reference to purpose. In this sense, the breeder's right is akin to that of the patentee's monopoly.

Article 14(2) makes provision for the scope of the breeder's right to extend to harvested material, including entire plants and parts of plants where these have been obtained through the unauthorized use of propagating material of a protected variety. However, the scope of the right is qualified, and exists unless: 'The breeder has had a reasonable opportunity to exercise his right in relation to the said propagating material'. The 1991 Act does not offer the breeder the choice between the exercise of his right in relation to the propagating material and the harvested

---

43   See Margaret Llewelyn, 'The Legal Protection of Biotechnological Inventions', *supra* at 118.

material. The right only extends to the harvested material where the breeder has no 'reasonable opportunity' to exercise his right over the propagating material.

The legal basis for the qualification is consent. A breeder presented with a reasonable opportunity of exercising his right in relation to propagating material, but who does not do so, is deemed to have consented to use of the material and is estopped in law from exercising his right in relation to harvested material resulting from the propagating material. Thus, the breeder is not totally free to exercise his intellectual property right over the grain instead of the seed. The burden of proving that the breeder had a reasonable opportunity to exercise his right over the propagating material rests with the alleged infringer. A commonly quoted example of the breeder being unable to exercise his right in relation to the propagating material is that of the piratical use of a breeder's variety in another country, perhaps a country which makes no provision for plant variety protection, followed by a subsequent import of harvested material of the variety into a country where the variety is protected.[44]

Article 14(3) provides for the further extension of the right of the breeder to products made from the harvested material. This provision is not, however, a part of the mandatory minimum scope of protection under the 1991 Act and states may choose whether or not to extend protection in accordance with Article 14(3). Under Article 14(3), authorization of the breeder is required to produce, sell or market[45] any product made directly from the harvested material, provided that the harvested material itself results from infringement. Again, the scope of the right is qualified, and exists unless: 'The breeder has had a reasonable opportunity to exercise his right in relation to the harvested material'.

Article 14 permits extension of the breeder's right to include mandatory protection of acts in respect of harvested material and optional protection for acts in respect of certain products of harvested material (subject to satisfying certain conditions). However, more is needed in awarding the breeder strong, patent-like, protection. By analogy with the absolute product protection principle in patent law, whereby all products, however made, receive protection once they result from a protected process, Article 14(3) should be a part of the mandatory minimum scope of protection under the revised UPOV; anything less is inadequate to protect the rights of the breeder.

Arguably, the most innovative feature of the 1991 Act is the introduction in Article 14(5) of the concept of 'essential derivation'. As already noted, under the 1978 Act[46] any variety was protectable provided it was clearly distinguishable from all other commonly known varieties in respect of one or more important characteristics. The Act further provided[47] that a protected variety could be used as an initial source of variation for the purpose of creating other varieties. Taken

---

44   See Barry Greengrass, 'The 1991 Act of the UPOV Convention', *supra* at 470.
45   And other acts referred to in Art. 14(1)(a), as outlined.
46   1978 Act Art. 6(1)(a).
47   1978 Act Art. 5(3).

together, the two provisions meant that the initial protected variety could be selected as a source of initial variation for the purpose of creating further variations. Selected in this manner, a variety was freely exploitable by the selector without any obligation to the breeder, as long as the selected variety was clearly distinguishable from the protected variety in respect of one or more important characteristics. Because 'important' was construed in the sense of making a distinction, a selector could select a mutant from an existing variety, or insert a gene into the variety by 'back-crossing' (a procedure in experimental genetics in which an offspring that is heterozygous at a locus or loci is mated with a homozygous individual) or some other procedure, and protect the resulting variety without rewarding the original breeder for his efforts. The 1991 Act seeks to remedy this situation by providing that a variety essentially derived from a protected variety cannot be exploited without authorization of the breeder of the protected variety.

Article 14(5)(b) states that a variety shall be deemed essentially derived from another variety (the initial variety) when:

- it is predominantly derived from the initial variety, or from a variety that is itself predominantly derived from the initial variety, while retaining the expression of the essential characteristics that result from the genotype or combination of genotypes of the initial variety;
- it is clearly distinguishable from the initial variety; and
- except for the differences which result from the act of derivation, it conforms to the initial variety in the expression of the essential characteristics that result from the genotype or combination of genotypes of the initial variety.

Article 14(5)(c) provides a non-exhaustive list of acts that may result in essential derivation, including the selection of a variant individual from plants of an initial variety, back-crossing, or transformation by genetic engineering.

The principle of essential derivation established under the 1991 Act highlights the similarities between the patent and plant variety rights systems and is intended to ensure that plant breeders and other biotechnologists cooperate with each other in the future, on any programme of activity being developed. An example of where cooperation is needed under the 1978 Act would be where, if a plant breeder inserts a patented gene into his variety, the resulting variety could fall within the scope of the patent. On the basis of the absolute product protection principle, whereby protection extends to a product obtained directly as a result of a patented process, the patentee can inhibit exploitation of the variety. By contrast, if the patentee inserts the patented gene into the same variety, the breeder has no possibility of preventing exploitation of the variety. Under the 1991 Act, if the patentee inserts his patented gene into a protected variety, the possibility exists that the modified variety would constitute an essentially derived variety and fall within the scope of protection. It is anticipated that the new balance established in this way between the two systems will facilitate exchange of technology between innovators.[48]

Because Article 14 of the 1991 Act awards strong, almost patent-like, protection to breeders of plant varieties, and is worrying for those who believe that such varieties should remain universally accessible for all to enjoy, it is imperative that safeguards in the public interest exist. However, adequate limitations are imposed in subsequent Articles ensuring that requirements of the public interest are complied with in the exercise of such rights. Viewed from this perspective, the plant variety rights system mirrors the patent system.

Public Interest and the UPOV 1991 Act

The interests of the public receive protection under the 1991 Act by virtue of the fact that:

- exceptions are imposed on the exercise of the breeder's right (Article 15);
- the breeder's right may become exhausted (Article 16), i.e., the right extends only to first, and not subsequent, sales, *unless* such sales involve further propagation of the variety;
- restrictions on the exercise of the breeder's right exist (Article 17).

Article 15(1) reiterates the so-called compulsory 'breeders' exemption'. It states:

> The breeder's right shall not extend to

> (iii) acts done for the purposes of breeding other varieties.

The provision reproduces the substance of Article 5(3) of the 1961 Act,[49] namely, authorization of the breeder is not required for use of a protected variety as an initial source of variation, if the purpose is to create other varieties, and is linked back to Article 14(5). The introduction of the principle of essential derivation in Article 14(5) does not necessarily mean a departure from the breeder's right exemption in that a variety is essentially derived only if it contains characteristics resulting from the genotype or combination of genotypes of the initial variety. The derived variety must resemble the initial variety in its genetic structure (to the full extent), subject only to the variation that may be expected from the particular features of its propagation. The concept of essential derivation is so specific that breeders will continue to be able freely to exploit newly bred varieties. In this manner, a balance exists between the interests of the breeder and those of the public. The principle of essential derivation, instead of diluting the breeders' exemption, may result in

---

48　See Barry Greengrass, 'The 1991 Act of the UPOV Convention', *supra* at 471.
49　As amended in the 1978 Act Art. 5(3).

strengthening the protection afforded to a protected variety, in that the genetic make-up of the initial variety is protected and is where the innovation lies.[50]

Other compulsory exceptions under the provision extend to acts done privately for non-commercial purposes,[51] and acts done for experimental purposes.[52] In this regard, the 1991 Act of UPOV is similar to certain sections of the Patents Act 1977 UK.[53]

An *optional* exception on the exercise of the breeder's right, in the form of the farmer's privilege, is created in Article 15(2). It reads partly thus:

> Each Contracting Party may, within reasonable limits and subject to safeguarding of the legitimate interests of the breeder, restrict the breeder's right in relation to any variety in order to permit farmers to use for propagating purposes, on their own holdings the product of the harvest which they have obtained by planting, on their own holdings, the protected variety.

In a number of respects, the language of the provision is limiting:

- the provision is discretionary: the breeder's right *may* be restricted;
- restriction may only occur if it is within *reasonable limits*;
- any restriction must take account of the *legitimate interests* of the breeder;
- breeders' rights can only be restricted in respect of *farmers*;
- the farmers must use the protected variety *on their own holdings*;
- restriction may only occur in respect of the *product of the harvest.*

The structure of the provision is such that states should give very careful consideration to the interests of plant breeders when exercising this option. The Diplomatic Conference responsible for the revision of UPOV formally recommended that the provision of Article 15(2) 'should not be read so as to be intended to open the possibility of extending the practice commonly called "farmer's privilege" to sectors of agricultural or horticultural production in which such a privilege is not a common practice'.

Article 15 of the 1991 Act of UPOV does not create a mandatory farmer's privilege but was inserted to permit states to continue national practice; however, states may restrict the *exercise* of the breeder's right if it is in the public interest to do so. By contrast, the patent system, also drafted to reflect states' practice, contains a mandatory exception to patentability in respect of the *grant* of the right by virtue of Article 53(b) EPC. This seems unnecessarily harsh on breeders.

---

50   See Barry Greengrass, 'The 1991 Act of the UPOV Convention', *supra* at 471.

51   1991 Act Art. 15(1)(i).

52   Ibid. Art. 15(1)(ii).

53   Section 60(5)(a) refers to acts done privately and for purposes which are not commercial, and s.60(5)(b) refers to acts done for experimental purposes relating to the subject-matter of the invention.

As already noted,[54] Article 53(b) EPC was inserted for reasons deferring to the UPOV system of protection. In the light of the revised UPOV, it makes no sense to continue the bar prohibiting the patenting of varieties contained in Article 53(b); a policy is required outlining a uniform approach to protection exceptions.

It is in the public interest that the breeder's right may become exhausted and Article 16 UPOV deals with the matter. Under Article 16, the breeder's right does not extend to acts concerning any material of the propagated variety which has been sold or otherwise marketed by the breeder or with his consent, unless such acts:

1. involve further propagation of the variety, or
2. involve an export of the material of the variety, which enables the propagation of the variety, into a country which does not protect varieties of the plant genus or species to which the variety belongs, except where the exported material is for final consumption purposes.

The intention underlying the provision suggests, once the breeder has consented to marketing the propagating material, he exhausts his rights in relation thereto. Arguably, however,[55] the effect of the provision is that the breeder's right to prohibit propagation of the variety is never exhausted; normally acts concerning the propagated variety will involve further propagation.

Article 17 UPOV deals with restrictions on the exercise of the breeder's right. Article 17(1) states that, except where expressly provided in the Convention, there can be no restriction on breeders' rights other than of 'public interest'. What does this mean? Are restrictions justified on grounds of public morality, public policy or public security? Given the communal importance of plant varieties, the public interest restriction outlined in Article 17(1) needs clarification.

Article 17(2) states that, where such restriction has the effect of authorizing a third party to perform any act for which the breeder's authorization is required, the Contracting Party shall take all measures necessary to ensure that the breeder receives 'equitable remuneration'. How is this to be measured?

The revised text of the UPOV Convention:

- provides more extensive protection for varieties than previous systems of protection;
- provides the same type and scope of protection as do patents;
- eliminates the bar on double protection; and
- protects adequately the interests of the public;

all of which suggests it can now be regarded as a viable alternative to the patent system.

54　See generally Ch. 2 heretofore.
55　See Barry Greengrass, 'The 1991 Act of the UPOV Convention', *supra* at 471.

Community Plant Variety Rights Scheme 1994

The provisions of UPOV 1991 have been substantially implemented in Community plant variety rights legislation. Council Regulation (EC) No. 2100/94 on Community Plant Variety Rights of 27 July 1994[56] (henceforth the CPVR) is the framework outlining protection for plant varieties within the EU.

Both the revised UPOV Convention and the CPVR scheme have led to a significant refinement of the plant variety right. The right now provides, not only more extensive protection for the plant breeder, but also a definition in more meaningful terms. By contrast, patentability criteria are not sufficiently defined under either the European Patent Convention 1973 or the Biotechnology Directive 1998.

The Recitals of the CPVR Regulation state that account is taken of international conventions such as the UPOV 1991 and the Convention on the Grant of European Patents 1973. The UPOV Convention was established to protect the interests of breeders and to promote the development of agriculture. By contrast, the Community scheme recognizes that industrial property regimes for plant varieties were, in the past, regulated by the legislation of Member States. At the Community level, in order to ensure the proper functioning of the European Single Market, harmonization of protection measures was necessary. In this regard, the policy underlying the two regimes differ fundamentally. Considering the objective of the Community scheme, it was appropriate that implementation, and application, be carried out by a Community Office; the Community Plant Variety Right Office was established.

The Community plant variety right is the sole and exclusive form of Community industrial property for plant varieties.[57] Nevertheless, the CPVR Regulation is expressed to be without prejudice to the right of Member States to grant national property rights for plant varieties.[58] It might be tempting, therefore, for Member States to protect plant varieties by means of a patent. However, the Regulation prohibits the subject-matter of a Community plant variety right from protection by either a national plant variety right or a patent.[59] The Regulation restricts the acquisition of cumulative protection over a single plant variety by both a patent and a plant variety right. Inconsistency between the intellectual property rights available for protecting plant varieties under UPOV and CPVR thus becomes apparent. Nevertheless, in many respects the scope of protection for breeders under UPOV and CPVR is similar and is outlined, briefly, as follows.

Article 13 CPVR deals with the rights of the holder and prohibited acts. Article 13(1) tells us that the holder shall be entitled to effect the acts set out in paragraph 2. The acts listed are the same as those set out in Article 14(1) UPOV[60] and include production and reproduction of variety constituents or harvested material

---

56 In force Apr 1995.
57 Regulation Art. 1.
58 Ibid., Art. 3.
59 Ibid. Art. 92(1).
60 *Supra.*

of the protected variety. Article 13(2) is without prejudice to Articles 15 and 16, which deal with limitation and exhaustion of rights, respectively. Furthermore, the holder may make authorization subject to conditions and limitations. Article 13(3) extends the rights of the holder to harvested material only if such material was obtained through the unauthorized use of variety constituents, and the holder had no reasonable opportunity to exercise his right in relation to the said variety constituents. The provision is equivalent to Article 14(2) UPOV.[61] Article 13(4) extends the rights of the holder. These include products directly obtained from material of the protected variety where such products were obtained through the unauthorized use of material of the protected variety and where the holder had no reasonable opportunity to exercise his rights in relation to the said material. The provision is equivalent to Article 14(3) UPOV.[62] Article 13(5) provides that paragraphs 1 to 4 of Article 13 shall also apply in relation to 'essentially derived' varieties. This provision is equivalent to Article 14(5) UPOV.[63]

CPVR and Morality

As already noted,[64] there is one important difference between CPVR 1994 and UPOV 1991, namely, the existence in the former of a morality provision absent in the latter. The purpose of the UPOV Convention is to ensure that Member States of the Union acknowledge the achievements of breeders of new plant varieties, by making available to them an exclusive property right on the basis of a set of uniform[65] and *clearly defined* principles. There is no morality provision under the Convention. Is this intentional? If so, it suggests that inclusion would not have resulted in uniform and clearly defined principles.

The Preamble to the 1961 Act of UPOV makes clear that special problems arise by virtue of the creation of the plant breeder's right. Hence, limitations that the requirements of the *public interest* may impose on the free exercise of such a right is permitted under the Convention. However, these problems are to be resolved by *each* state in accordance with uniform and clearly defined principles.[66] By contrast, harmonization within the EU obliges all Member States to operate the same system of law, and resolution of problems by *each* is not possible. As already noted, the CPVR Regulation takes into account international conventions such as UPOV and the EPC. In addition, the recitals of the Regulation highlight the need to conform to existing (largely European) procedures. A clear example is that it is not intended to alter definitions which have been established in the patent field.[67]

---

61   *Supra.*
62   *Supra.*
63   *Supra.*
64   *Supra.*
65   Regulation Recital 6.
66   1961 Act of the UPOV Convention Preamble at para. 3.
67   Regulation Recital 13.

The definition of plant variety is the same under CPVR as it is for patent law purposes.[68] Whether, and to what extent, the conditions for protection accorded in the patent system should be adapted or otherwise modified for consistency with the CPVR is uncertain.[69] In order to limit the extent of protection, the *exercise* of Community plant variety rights is subject to restrictions laid down in provisions adopted in the *public interest*.[70] In this manner, the CPVR scheme is safeguarding the interests of agriculture while at the same time protecting those of breeders.

The morality provision is contained in Article 13(8) CPVR and provides that the *exercise* of the rights conferred by a Community plant variety right may not violate *any provisions* adopted on the grounds of '[p]ublic morality, public policy or public security, the protection of health and life of humans, animals or plants, the protection of the environment, the protection of industrial or commercial property, or the safeguarding of competition, of trade or of agricultural production'. Article 13(8) is without prejudice to Article 14 and Article 29, which deal with derogation and compulsory exploitation of Community plant variety rights respectively. Article 13(8) refers to both Community and national provisions, meaning that a Community plant variety right, once granted, does not permit the rights holder to use the right for any purpose whatsoever, but that Member States may restrict the exercise of the right by national legislation. Instead of determining morality by reference to grant, determination over the exercise of the right once granted suggests moving the question of morality away from the intellectual property right per se. This is in stark contrast to the operation of the patent system where a determination of morality lies in respect of the *grant* of the right itself[71] and the question of morality impinges upon the *existence* of the intellectual property right.

Public Interest and CPVR

The interests of the public are adequately protected under CPVR; derogation, limitation and exhaustion of rights exist by virtue of Articles 14, 15 and 16 respectively.

Derogation is provided for in Article 14 and creates, in the public interest, a mandatory, albeit limited, farmer's privilege. The reason why a mandatory farmer's privilege exists under CPVR is due to the special position which agriculture holds within the Community. Regulation No. 2100/94 was drafted to reflect states' practice[72] and permit farmers certain privileges for the purposes of promoting agriculture. The Recitals of the Regulation state that the effect of a Community

---

68   Ibid.

69   Ibid. Recital 29.

70   Ibid. Recital 21.

71   EPC Art. 53(a) says that a patent shall be refused where publication (of the application) would be contrary to *ordre public* or morality.

72   In this regard, see the Patents Act 1977 amended by the Patents Regulations 2000 Schedule A 1.

plant variety right is uniform throughout the Community, and commercial transactions subject to the holder's agreement are precisely delimited. Farmers are permitted under Article 14(1) to use, for propagating purposes in the field on their own holding, the product of the harvest obtained by planting on their own holding, propagating material of a variety other than a hybrid or synthetic variety, covered by a Community plant variety right.

Article 14(3) sets down the derogation criteria and include:

- Small farmers[73] shall not be required to pay any remuneration to the holder.
- Other farmers shall be required to pay equitable remuneration to the holder, which shall be sensibly lower than the amount charged for the licensed production of propagating material of the same variety in the same area.

The Community Regulation permits equitable remuneration only by way of an amount 'sensibly lower' than the licence to be collected in respect of certain species. What the term 'sensibly lower' means, and how it can be reconciled with the word 'equitable', has not yet been resolved.

The public interest is further protected by limiting the breeder's Community rights and is dealt with in Article 15 CPVR. It states that CPVR shall not extend to:

- acts done privately for non-commercial purposes;[74]
- acts done for experimental purposes;[75]
- acts done for the purpose of breeding, or discovering and developing other varieties.

In conformity with UPOV, authorization of the breeder is not required for use of a protected variety as an initial source of variation for the purposes of creating other varieties, except where the other variety is an 'essentially derived' variety.[76]

Exhaustion of breeders' Community rights is dealt with in Article 16. Community plant variety rights shall not extend to acts concerning any material of the protected variety, including essentially derived varieties, which has been disposed of to others by the holder or with his consent, in any part of the Community, or any material derived from the said material. However, the right does extend to acts relating to the further propagation of the variety, unless such propagation was intended when

---

73   Farmers who do not grow plants on an area bigger than the area which would be needed to produce 92 tonnes of cereals in the case of those plant species to which Council Regulation (EEC) No. 1765/92 of 30 June 1992 applies. See CPVR 1994 Art. 14(3) at para. 5.

74   Similar to the Patents Act 1977 s 60(5)(a).

75   Ibid. at s.60(5)(b).

76   Regulation Art. 15(d).

the material was disposed of.[77] The right also extends to acts involving export of variety constituents into a third country and does not provide variety protection, unless the exported material is for final consumption purposes.[78]

To ensure that varieties remain universally accessible for all to enjoy, the possibility exists for compulsory exploitation of the rights. Compulsory exploitation rights are granted on application to the Community Office on grounds of 'public interest' only, as set out in Article 29. The Office shall stipulate the types of acts covered and specify reasonable conditions pertaining to the compulsory exploitation right. Such reasonable conditions shall include the interests of the holder of the plant variety right, and may include a possible time limitation. It may also include the payment of an appropriate royalty as equitable remuneration to the holder. Certain obligations may be imposed on the holder in order to make use of the compulsory exploitation right.[79] On the expiry of each one-year period after the grant of the compulsory exploitation right, parties to the proceedings may request that the decision on the grant be cancelled or amended.[80] The compulsory exploitation right shall also be granted to the holder in respect of an essentially derived variety, if it is in the public interest to do so.[81] In respect of Community plant variety rights,[82] compulsory exploitation is not granted by Member States. This is significant in that, on the one hand, Article 29 permits a patent holder to claim a compulsory licence in respect of a protected plant variety, and control is by the CPVR scheme. On the other hand, the CPVR scheme does not permit a holder to claim a compulsory licence in respect of material protected by a patent.

The morality clause, Article 13(8) CPVR, is expressed without prejudice to the compulsory exploitation provision set out in Article 29, suggesting that the term 'public interest' is not equated with 'public morality'. If so, the answer to the question, namely, whether or not 'public interest' under Article 17 UPOV incorporates public morality, is likely to be no.

Most of the principles underlying UPOV are incorporated into CPVR. The reason is that many of the problems involved in protecting breeders were resolved at the UPOV revision stage. Because formulations exist, the European Commission, by aligning the Biotechnology Directive more closely with existing Community legislation, could have achieved uniform measures of protection for life forms under a single regime. However, because the Directive restates EPC principles, this suggests EPO jurisprudence is likely to persist, thus perpetuating a dual system of protection for living material.

---

77   Ibid. Art. 16(a).

78   Ibid. Art. 16(b).

79   Ibid. Art. 29(3).

80   Ibid. Art. 29(4).

81   Ibid. Art. 29(5), and the Administrative Council must also be consulted for its advice.

82   Ibid. Art. 29(7).

Directive on the Legal Protection of Biotechnological Inventions 1998

The Directive on the Legal Protection of Biotechnological Inventions 1998 aims to provide the framework whereby inventions based on living organisms receive protection within the EU. In adopting the Directive, the European Commission sought to prevent a patchwork of national laws on intellectual property rights which could hamper a single market in biotechnology products.[83] However, what the Directive demonstrates is that the Commission believes there is a need for additional legislation to supplement the EPC. The Directive purports to extend patent protection to biological material,[84] meaning any material containing genetic information capable of reproducing itself or being reproduced in a biological system. Critically, the Directive mentions specifically that plant and animal varieties are not patentable.[85] Is the reason for exclusion that varieties are not biological material, or because plant variety material receives protection by virtue of the CPVR regime? If the latter, it suggests that an animal variety rights scheme exists and it does not, and clearly, varieties are biological material. Therefore, either the Biotechnology Directive is deferring to the UPOV system of protection, which makes no sense since the bar on double protection is there eliminated, or to states' practice. Whatever the reasons for exclusion, the Biotechnology Directive continues the ban contained in Article 53(b) EPC, suggesting that the Directive will be governed by EPO jurisprudence. However, because the EPO has confirmed that inventions consisting of living material are patentable, what the Directive adds in terms of protection deserves consideration. Notwithstanding the ban on patenting plant and animal varieties, the Directive permits patenting of inventions concerning plants or animals if the technical feasibility is not confined to a particular plant or animal variety.[86]

Both UPOV 1991 and CPVR 1994 bear a direct correlation to the objectives and provisions of the Biotechnology Directive 1998. Plants consist of self-replicating biological material and the Directive offers the same scope of protection to innovators of plant technology as is available to breeders under Article 13 CPVR and Article 14 UPOV. Turning away from the confines of the patent system indicates that existing EPO jurisprudence is insufficient for protecting living material. As already noted, restating, rather than reformulating, existing patent law principles in the Directive overlooks the need for consensus between the different systems available for protecting living material. To ensure consistency with its own existing legislation, the Commission should have looked to the CPVR system.

---

83   In this context, see the Patents Act 1977 amended by the Patents Regulations 2000.

84   Directive Article 2(1)(a).

85   Ibid. Art. 4(1).

86   Ibid. Art. 4(2).

Scope of Protection

The Directive, in Article 8, recognizes the well-established patent law principle that protection extends to a product, process or product obtained directly as a result of a patented process. In affording patent protection to 'second'- and subsequent generation products, the Directive recognizes special features of inventions containing biological material, namely, that breeding processes are not descriptively repeatable to order. The Directive recognizes that adequate protection of biological material necessitates extension to future generations. In this sense, the Directive is akin to the CPVR scheme. The Directive, in Article 8(1), recognizes that protection conferred by a patent on biological material possessing specific characteristics shall extend to any biological material derived from that biological material through propagation or multiplication and possessing the same characteristics. Article 8(2) recognizes that a process protected by patent, and enabling production of biological material possessing specific characteristics, extends to biological material obtained directly through the process, and to any other biological material derived from the directly obtained biological material through propagation or multiplication and possessing the same characteristics. Because the scope of protection under the Directive and CPVR is similar suggests that:

- plant material is biological and exclusion of varieties under the Directive on this ground is not justified, and
- distinguishing between plants and animals is not always clear and exclusion under the Directive on the basis that plant material receives protection by virtue of CPVR is difficult to justify.

Derogation of Rights and the Public Interest

Under the Directive the interests of the public are adequately met by permitting derogation and compulsory cross-licensing of rights. Article 11 contains a mandatory *farmer's privilege* exception and in this respect is similar to Article 14 CPVR. Article 11(1) creates the farmer's privilege and reads as follows:

> By way of derogation from Articles 8 and 9, the sale or other form of commercialisation of plant propagating material to a farmer by the holder of the patent or with his consent for agricultural use implies authorisation for the farmer to use the product of his harvest for propagation or multiplication by him on his own farm, the extent and conditions of this derogation corresponding to those under Article 14 of Regulation (EC) No. 2100/94.

This has been met by criticism that continuation of 'privilege' dilutes protection granted under the Directive, and even runs contrary to the patent system as a whole. However, such criticism is not well founded.

Article 11(1) implies that the 'privilege' extends from year to year and in this manner subsequent use of material derived from a protected plant is permitted. The approach adopted in the Directive reflects states' practice and is welcome because the practice of farmers retaining material from one year to the next would in any event, be a difficult one to prevent.[87] Even if the Directive expressly stipulated[88] that no *subsequent* use of material derived from a protected plant is permitted,[89] the difficult question of 'enforcement' must be considered. It is likely that the farmers' practice of retaining seed will continue, and, as such, attempts in the Directive to deal with the issue are welcome. Imposing the strict patent ideal of absolute monopoly in such circumstances is not helpful and likely to have the effect of alienating a farming community already suspicious of the reasons why patent protection over crops and fodder material is necessary. Two comments can be made. First, the farming community is not experienced in patent law principles. Second, the farming community finds it difficult to see how the patent system has a direct application in the context of farming.

Arguments over the inclusion of the farmer's privilege highlight a distinction between the patent and the plant variety right systems.[90] Because the patent system is not industry-specific, it can be seen as separate from those who make use of patented products and processes. The focus of attention is on the rights of the patent holder rather than the interests of the user of the protected material. By contrast, in the Community plant variety rights scheme there is a close relationship between the system, its methods of operation, and those who use the protected varieties. Without the support of the farming community, many protected varieties would cease to have any value, as they would be ignored in preference for other crop material. The reactions of farmers to varieties grown are crucial to decisions of plant breeders about new breeding programmes. Consensus between the interests of the breeder and those of the farmer is regarded as vital. Is the Community scheme an accepted form of legal protection for living material in a way that the patent system is not?

In regard to farmer's privilege, the extent and conditions of derogation correspond to those under Article 14 of Regulation (EC) No. 2100/94. The Community Regulation permits an equitable remuneration by way of an amount 'sensibly lower' than the licence to be collected in respect of certain named species. The wording of the Directive suggests that equitable remuneration lies in respect of specifically mentioned plant species only. In linking derogation under the Directive with Community plant variety rights under the Regulation, the Commission has

---

87   See Margaret Llewelyn, 'The Legal Protection of Biotechnological Inventions', *supra* at 125.

88   However, it does not.

89   Including where the farmer uses it for his own purposes on his own land.

90   See Margaret Llewelyn, 'The Legal Protection of Biotechnological Inventions', *supra* at 125.

recognized that the latter affords strong, almost patent-like, protection. This suggests CPVR legislation could have better informed the Directive.

Other anomalies under the Directive include a breeder's privilege in respect of animal reproductive material only. Article 11(2) creates the breeder's privilege and reads as follows:

> By way of derogation from Articles 8 and 9, the sale or any other form of commercialisation of breeding stock or other animal reproductive material to a farmer by the holder of the patent or with his consent implies authorisation for the farmer to use the protected livestock for an agricultural purpose. This includes making the animal or other animal reproductive material available for the purposes of pursuing his agricultural activity but not sale within the framework or for the purpose of a commercial reproduction activity.

Derogation does not extend to plant reproductive material. Is the reason that plant reproductive material is covered by the CPVR regime? It seems anomalous that, on the one hand, the Directive incorporates derogation provisions relating to the CPVR scheme while, on the other, denying that such provisions extend to plant reproductive material.

Article 11(3) governs the extent and conditions of derogation, and is to be determined by national laws, regulations and practices. This suggests that, subject to the Directive, Member States can impose controls on the right of the farmer to use protected livestock for his own purposes. The extent of such restrictions may or may not follow the same lines as for plant variety derogation. Under Article 11(3) of the Directive there is no requirement that equitable remuneration must be paid to the patentee before derogation under Article 11(2) is permitted. It is possible that Member States could, if they wish, altogether ignore Article 11(2) and provide no derogation at all. Such a situation would clearly not be in the public interest[91] and would allow the patentee to claim an amount equivalent to the royalty he expected to receive if the farmer had bought the progeny from the patent holder, instead of breeding it for himself.

It may be unsuitable to permit Member States to make independent determination of the extent to which derogation should apply in respect of livestock. Consider, for example, the situation where an animal is sold from one Member State to another. The 'state of sale' may permit only a minimal derogation from the rights of the patent holder who is domiciled there, whereas the 'state of purchase' (where progeny are likely to be born) may permit a more extensive derogation. How are such situations to be resolved? In this respect, the Directive does not give guidance.

The interests of the public are also upheld by virtue of Article 12 of the Directive in which the grant of a compulsory cross-licence in respect of both the

---

91   In this context, the Directive is lacking.

breeder and the patent holder is permitted. Article 12(1) deals with the breeder, and reads partly as follows:

> Where a breeder cannot acquire or exploit a plant variety right without infringing a prior patent, he may apply for a compulsory licence for non-exclusive use of the invention protected by the patent in as much as the licence is necessary for the exploitation of the plant variety to be protected, subject to payment of an appropriate royalty.[w]here such a licence is granted, the holder of the patent will be entitled to a cross-licence on reasonable terms to use the protected variety.

Article 12(2) deals with the patent holder, and reads partly as follows:

> Where the holder of a patent concerning a biotechnological invention cannot exploit it without infringing a prior plant variety right, he may apply for a compulsory licence for non-exclusive use of the plant variety protected by that right, subject to payment of an appropriate royalty.[w]here such a licence is granted, the holder of the variety right will be entitled to a cross-licence on reasonable terms to use the protected invention.

In respect of both Article 12(1) and Article 12(2), the right to claim a compulsory licence is qualified in three respects:

- the applicant must have applied unsuccessfully to the other rights holder to obtain a contractual licence;
- the plant variety or invention must constitute significant technical progress;
- the plant variety or invention must be of considerable economic interest.

The qualifications are laid down in Article 12(3)(a) and (b).

The principle of mutuality embedded in the Directive suggests that plant variety rights protection is now patent-like. On this basis, it is anticipated transfer of technology between the breeder and inventor will be facilitated, for example by the compulsory cross-licensing provision, and is welcome. Because a compulsory licence is granted only where 'considerable economic interest' is shown and, if 'considerable economic interest' has a meaning similar to 'public interest', it is difficult to see grounds for objection.[92] And the fact that licensing of either the invention or the plant variety takes place only where it is shown that the invention or plant variety concerned constitutes 'significant technical progress' suggests it is unlikely that the provision will be invoked on a regular basis. Nevertheless, although the provision is welcome, it is not beyond criticism. On the one hand, a compulsory licence in respect of a plant variety right may be granted by the Community Office in accordance with Article 29(7) of Regulation

---

92   In this context, the Directive does not give guidance.

2100/94. On the other, a compulsory licence in respect of a patent is granted by the designated authority in each Member State.[93] On this basis, the possibility exists that a divergence in practice between Member States will emerge in respect of determining both what amounts to an 'appropriate royalty', and conditions of grant, for the compulsory licence.

One aim of the European Commission is to ensure the proper functioning of the internal market. If the patent system as currently operating within the EU is an obstacle to achieving that aim, and the implementation of the Directive suggests this, a more appropriate system of protection is required. What is required is consistency with existing European legislation in the form of the CPVR scheme.

Conclusions

Under the revised UPOV 1991, exclusive property rights are made available to breeders on a set of uniform and clearly defined principles, namely, distinctness, uniformity and stability. The scope of protection, farmer's privilege and cross-licensing are issues discussed and already resolved there. By contrast, within the patent system the issues of inventive step, obviousness and who is the man skilled in the art remain unresolved, suggesting that protection there is not awarded on clearly defined principles.

Both the revised UPOV Convention and the CPVR scheme have led to a significant refinement of the plant variety right. Under CPVR the right is now defined in more meaningful terms, suggesting it is an *accepted* form of protection for living matter, in a way that the patent system is not. The right, once granted, can be subjected to other restrictions, thereby controlling exploitation of the protected material. The type and scope of protection available under plant variety rights schemes suggest there is a viable alternative to the patent system.

The UPOV system for protecting plant varieties is without a morality provision. The CPVR scheme contains a morality provision exercisable in respect of the *use* of the right only. Under the CPVR scheme, issues relating to morality are removed from the framework of the intellectual property rights system. This is a sensible way of dealing with the issue of morality. By contrast, the morality provision of the Directive is exercisable in respect of the *grant* of the right. In this respect, the Directive runs counter to international trends and could result in confusion in respect of states' moral obligations.

The mere existence of the Biotechnology Directive demonstrates that the Commission believes there is a need for additional legislation to supplement the EPC. Since existing EPC provisions, in particular Article 53(b), deny protection to innovators of plant technology, the Biotechnology Directive should have reformulated rather than simply restated such principles. In this regard, the underlying premise of the Directive, namely, that it must comply to a greater

---

93   Directive Art. 12(4).

rather than lesser extent with existing provisions of the EPC, is questionable. The Biotechnology Directive overlooked the need for consensus between the different systems available for protecting living material. In drafting the Biotechnology Directive, the Commission should have looked to its own existing form of protection for guidance, namely, the CPVR scheme, thus, as a minimum, ensuring, in the absence of a single regime of protection, consistency of practice between the two systems of protection.

In its present form, the Biotechnology Directive is likely to continue the controversy surrounding the patenting of, in particular, living plant material.

Chapter 9
Recent EU Initiatives:
Likely Impact on Moral Concerns

Introduction

Biotechnology has been part of our heritage since the dawn of civilization. However, biotechnology has been of concern to patent lawyers only since the novel use of organisms in the context of DNA structure was discovered.[1] The nature of the subject-matter gives rise to complex conceptual, theoretical, moral and environmental questions in regard to the application of the exclusions to patentability of biotechnological inventions.

Law and Morality

In European states under the European Patent Convention and the Directive on the Legal Protection of Biotechnological Inventions 1998, the relationship between law and morality has been an uneasy one.[2] Support for this is found in two situations: the decisions of the European Patent Office on patentability, and the difficulties encountered in adopting the Directive.

The EPO is unable to offer guidance as to the meaning and application of Article 53(a) EPC and that jurisprudence is unclear in respect of what is unpatentable by virtue of it. However, what EPO jurisprudence does make clear is that morality is now a matter for European institutions. This suggests EPO jurisdiction has expanded beyond that envisaged in the EPC by virtue of the creation of Article 53(a). Article 53(a) was created to continue national practice in respect of powers to refuse patents on the basis of morality. In effect, only those inventions concerned with sexual morality were precluded from patentability. Because sexual morality is no longer an issue in European states under the EPC, this suggests morality per se has a residual role in the patent system. Likewise, EPO jurisprudence is unclear in respect of what is unpatentable by virtue of Article 53(b) EPC. Because EPO jurisprudence is unclear in respect of subject-matter patentability, Article 53(b) has resulted in inconsistency of practice among EPC states.

---

1   See Anthony McInerney, 'Biotechnology: *Biogen v Medeva* in the House of Lords', 1998 *European Intellectual Property Review* 14–21 at 14.

2   See Lionel Bently and Brad Sherman, 'The Ethics of Patenting: Towards a Transgenic Patent System', 1995 vol. 3 *Medical Law Review* 275–91 at 277.

The second situation which illustrates the law's discomfort when dealing with moral concerns is in relation to the Directive. The Directive in its initial form has the distinction of being the first directive ever rejected by the European Parliament. A function of the Directive is to clarify the EPC in relation to the patenting of genetic material and to facilitate a uniform application of Article 53 in Member States. The original proposal made no reference to moral considerations. However, by the time of its adoption in 1998 moral concerns informed much of the Directive. Yet, despite formal agreement, there remain deep divisions of opinion. The weakening of earlier opposition to the Directive reflects a practical accommodation with the commercial importance of patenting biotechnological inventions.

Widening the Debate

Within the patent system the internal debates reflect the rationalities of the three principal groups of participants: those of law, science and economics. But, even for these groups, the appropriate application and development of legal terms to scientific practices, and the question of whether the practices of the courts and patent offices adequately meet the economic goal of providing incentives for invention and commercial development is problematic.[3] The advent of biotechnology has posed significant challenges for patent law. As in any legal system, there are areas of contention and uncertainty as to the application of legal provisions to particular fact situations.[4] Interpretation of the provision which excludes plant and animal varieties from patentability,[5] the meaning of non-obviousness and inventiveness,[6] and whether or not patents are being granted that are too wide in their scope,[7] are examples of uncertainty in the context of biotechnology, suggesting other criteria on the margins are not helpful.

Likewise, outside the patent system, sources of controversy exist. The granting of patents over plant, animal or human material is opposed by many[8] who believe that exploitation of such living matter is not appropriate. Whether or not this is no more than an indirect way of impugning biotechnology without a reason is difficult to determine. In particular, patenting of human material in the form of gene sequences is considered to be wrong because this amounts to the commercialization of life. However, DNA is not 'life' but a chemical substance carrying genetic information.[9] Can the claim that patenting higher life forms is

---

3   The experiences in the UK of Genentech and Biogen have caused much in the way of speculation about the future relationship between the industry and patent offices.

4   See Julia Black, 'Regulation as Facilitation: Negotiating the Genetic Revolution', 1998 vol. 61 *Modern Law Review* 621–60 at 646.

5   See generally Ch. 4 heretofore.

6   See generally Ch. 5 heretofore.

7   See *Biogen*, per Lord Hoffmann, discussed generally in Ch. 5 heretofore.

8   For example Greenpeace, Friends of the Earth and religious groups.

9   See *Howard Florey/Relaxin* [1995] EPOR 541, discussed in Ch. 7 heretofore.

intrinsically wrong and morally reprehensible be justified on the basis that human beings have discovered how to change and alter biological evolution which, before, appeared to be totally beyond the realm of human control and in the sphere of natural or divine forces only? On the one hand, genetic engineering does raise questions about the limits and possibilities of human control over life. On the other, while the prospect of this shift in control may be disquieting, it cannot be morally wrong to exercise it. Support for this is found in *Plant Genetic Systems* where the Technical Board recognized that development of this technology inevitably allows a better understanding and control of the natural phenomena linked to plants, but this did not render activities in this technical field intrinsically wrong.[10] This argument is also reinforced by the fact that humans have been exercising control over 'life' for some considerable time past.[11]

Additionally, patents are being objected to on the grounds, not only that one person should be able to conduct the activity for which the patent is sought, but also that *no one* should. In this regard, the Directive expressly provides,[12] *inter alia*, that processes for modifying the germ-line genetic identity of human beings are unpatentable. This is not because such activities should be open to all to enjoy; rather it is felt that these activities are morally insupportable.

Governments must address difficult and complex issues with the right information, understanding and analysis.[13] In this regard, a number of key challenges exist:[14]

- how to make complex issues more understood by the general public;
- how to ensure that advice reflects the broad range of opinion with the correct analysis; and
- how to ensure that the systems that emerge to address issues in biotechnology are worthy of public confidence.

The Directive, in not addressing adequately the moral concerns associated with biotechnology and methods of risk assessment, has helped to ensure that the attitude of the public is unlikely to change, or will only change slowly.[15] Additionally, the Directive, drafted with the underlying idea that patentability requirements should be kept in conformity with the EPC and its current interpretation at the EPO, has, in respect of biotechnological inventions, analysed the situation incorrectly.

---

10   *Plant Genetic Systems* [1995] EPOR 357 at point 17 of the Reasons for the Decision.

11   Examples include birth-control techniques and insulin treatment for diabetics. See generally Ch. 1 heretofore.

12   Article 6(2)(d).

13   Colin Campbell, 'A Commission for the 21st Century', 1998 vol. 61 *Modern Law Review* 598–602 at 599.

14   Ibid.

15   See generally Ch. 7 heretofore.

Harmonization can be achieved only when Member States of the EU either reach a consensus as to the meaning of morality within the context of patentability, or reject that approach altogether. Because the Directive does not inform the public sufficiently as to the moral problems involved in biotechnology, this suggests that it may not be worthy of public confidence. On these bases, the effect of the Directive is weakened.

Critics of genetic engineering argue that the introduction of commercial norms is inappropriate in certain circumstances. In essence, the objection is that biotechnology in general, and genetic engineering in particular, is a risky business. Genetic engineering is wrong as a result of its potential deleterious consequences. However, in response, development of all new technology is accompanied by risks.[16] Critics also argue that the benefits which genetic engineering may produce are outweighed by the harms it is likely to cause. What are the relevant benefits and harms? In the context of genetic engineering and the creation of transgenic species, animal suffering is a harm that is relevant.[17] Likewise, release into the general environment of genetically modified organisms which could cause irreversible adverse effects is a relevant harm.[18] But can the benefits and harms be assessed? The criteria include weighing the suffering to animals and possible risks to the environment on the one hand, and the invention's usefulness to mankind on the other.[19] How are such criteria assessed? There is no suggestion in EPO jurisprudence of any method for assessing such criteria. How probable is the occurrence of harm? The mere fact that uncontrollable risks are conceivable cannot be a major factor in determining whether or not a patent should be granted.[20] Additionally, for every technology the most profound consequences, positive or negative, are the ones that are *at present unforeseeable* and for that reason impossible to include in the comparison.[21] On this basis, the consequential arguments are suspect.

US Patent System: A Model?

The differences that exist between patent systems worldwide suggest harmonization of laws on an international level is required. In the past, before the advent of the EPC, this was difficult for European states as views of national legislators differed on concepts of patent protection, patent law as a matter of economic policy and patent law as an integral part of the general legal system of the country. And, more

---

16   See *Harvard/Onco-Mouse* [1991] EPOR 525 discussed generally in Ch. 4 heretofore.

17   Ibid.

18   Ibid.

19   Ibid.

20   Ibid.

21   See *Plant Genetic Systems* [1995] EPOR 357 at point 18.5 of the Reasons for the Decision.

recently, in this regard the experiences of the European Commission in introducing the Directive are not encouraging either, although these issues arose by virtue of problems of a moral rather than a practical nature.

Commentators such as Armitage contend that it should be an aim of intellectual property policy in the EU, as a crucial element of the EU's policy for industry, that European industry enjoys as good protection by industrial property rights as its American competitors.[22] As far as patents are concerned, it is clear that European states under the EPC suffer compared with the US. Armitage argues that patent law in the EU needs to be as closely under Community control as US patent law is under federal control if it is to be as capable of adjusting to changing circumstances, particularly developing technology.[23] Is Armitage correct in his analysis?

In the US the requisite control is achieved via the legislative machinery of the country and by government administration. In the EU, the equivalent legislative processes are Council regulations and directives, and the equivalent administrative control is by the European Commission. As regards legislation, the Community cannot be said to be master of its own destiny. Its patent law is determined by the EPC, which is not restricted in membership to Community countries, and the Directive, which is. As regards administration, the EPO is not subject to control by the Community in respect of either its finances or procedures. On this basis, the views of Armitage deserve careful consideration.

Because the US Constitution permits Congress to enact patent legislation, reform is facilitated. Legislative reform in the US has resulted in the creation of the Court of Appeals for the Federal Circuit with exclusive jurisdiction over all appeals in patent cases. European patents suffer compared with US patents in that infringement actions have to be taken country by country with no certainty of consistent results. Other legislative reforms by Congress include extension of patent protection to a broad range of biotechnological inventions such as plant technology,[24] meaning that plant material does meet the criteria for patentability. Hence, in respect of life form protection a single regime operates in the US. Additionally, in the US the judiciary has embarked on judicial legislation and the courts there have laid down the conditions for patentability. In this regard, it was never intended morality should extend to validity criteria. Because of this, courts in the US do not have to consider whether, and to what extent, moral norms change with time. Neither do US courts have to deal with the limits of the morality test.[25] The US Patent and Trademark Office (henceforth PTO) functions as a patent-granting authority and not a determiner of patent policy. However, in this regard, problems exist. Applicants to the PTO complain about the PTO backlog, the PTO

22　See Edward Armitage, 'EU Industrial Property Policy: Priority for Patents?', 1996 vol. 10 *European Intellectual Property Review* 555–8 at 555.

23　Ibid.

24　See generally Ch. 6 heretofore.

25　Whereas there are no limits on the morality test in the EU, risk assessment is high on the agenda.

complains about low-quality patent applications, and industries complain about patents that, some believe, should not have been granted.

In the past, the PTO, Congress and the courts all have taken steps to reform the US patent system. But, in light of recent events, meaningful patent reform via the PTO remains uncertain. A US district court recently blocked the PTO's major attempt at reform. Now, interested parties are looking to Congress and its debate on the Patent Reform Act 2007. In January 2006 the PTO proposed significant rule changes in order to address the backlog of unexamined patent applications. The Rules made changes to patent prosecution in several areas including, limiting the number of claims that can be presented without having to provide a detailed patentability analysis. According to the PTO, the Rules were to be effective 1 November 2007. However, just after the Rules were published, a sole inventor[26] challenged their validity in district court.[27] Soon after, SmithKline Beecham brought a similar suit, and the two cases were joined. Both parties sought a preliminary injunction to stop the Rules from taking effect. The injunction was granted, the court finding that the parties were likely to succeed in their challenge to the Rules. On 1 April 2008 the district court issued a final opinion, and order, invalidating the Rules, and enjoined the PTO from implementing them.

The court's decision is important on two fronts. First, it blocked the Rules from taking effect. Second, it held that the PTO does not have substantive rule-making authority. The PTO is now expected to ask Congress to legislatively reverse its district court defeat. It seems that in the US patent reform will continue to be a 'hot' topic for several more years.

EU Initiatives

The Biotechnology Directive 1998 tells[28] that inventions shall be considered unpatentable where their commercial *exploitation*[29] *would be contrary to ordre public* or morality. This is an attempt by the EU to confine morality to commercial situations and is welcome. However, are the concepts of *ordre public* and morality clear? The Directive sets out which inventions involving plants, animals or the human body may or may not be patented. It requires Member States to allow the patenting, under certain conditions, of inventions which may have industrial application making it possible to produce, process or use biological material.

Whether or not the Biotechnology Directive facilitates harmonization of the laws of Member States was considered in *Netherlands v Parliament and Council*[30]

---

26   Triantafyllos Tafas.
27   Eastern District of Virginia.
28   See Art. 6.
29   My emphasis.
30   Case C-377/98, decision of 9 Oct 2001. Italy and Norway supported the position of the Netherlands.

where the European Court of Justice (ECJ) took the view that the Directive enhanced
the smooth operation of the internal market because it helps to bring the laws of
Member States into line and to remove the legal obstacles to the development of
activities in the field of genetic engineering. The Netherlands complaint that the
concepts of *ordre public* and morality used by the Directive are unclear, or even
contradictory, was rejected by the ECJ. The Netherlands' chief argument was that
the patentability of isolated parts of the human body provided for by Article 5(2)
of the Directive reduces living matter to a means to an end, undermining human
dignity. The ECJ noted that:

1. respect for human dignity was provided for in Article 5(1) of the Directive
   which provides that the human body at the various stages of its formation
   and development cannot constitute a patentable invention;[31]
2. the elements of the human body are not patentable in themselves and their
   discovery cannot be the subject of protection.[32] An element of the human
   body may be part of a product which is patentable but it may not, in its
   natural environment, be appropriated;[33]
3. additional security is offered in Article 6 of the Directive, which cites
   as contrary to *ordre public* and morality, and therefore excluded from
   patentability, processes for cloning human beings, processes for modifying
   the germ-line genetic identity of human beings and uses of human embryos
   for industrial or commercial purposes;[34]
4. it is clear from those provisions that, as regards living matter of human
   origin, the Directive frames the law on patents in a manner sufficiently
   rigorous to ensure that the human body effectively remains unavailable and
   inalienable and that human dignity is thus safeguarded.[35]

Because the validity of the Directive has been upheld in this manner suggests there
is unlikely to be a legislative return to issues, including moral ones, associated
with biotechnology, for a long time.

Outline of Developments

When it adopted a First Action Plan for Innovation in Europe,[36] the Commission
deemed it essential to gain as full a picture as possible of the patent system in
Europe in order to: assess whether it meets the needs of users; examine whether
new Community measures are necessary and, if yes, what form they should take.

---

31 See the judgment at point 71.
32 Ibid. at 72.
33 Ibid. at 73.
34 Ibid. at 76.
35 Ibid. at 77.
36 COM (96) 589 final, 20 Nov 1996.

This resulted in the *Green Paper on the Community Patent System in Europe.*[37] The European patent system is based on two international agreements, the EPC 1973 and the Community Patent Convention 1975, which is an integral part of the Agreement relating to Community patents 1989.[38] The EPC does not create a uniform protection right. What is more, there is no provision within this system for a court with powers to settle patent disputes at European level. The Community patent is intended to bring together the bundle of protection rights resulting from the grant of a European patent and merge them into a single, unitary protection right valid throughout the Community. Under the EPC, the management of proceedings for infringement or revocation is complex and has to be brought before the national courts of each country for which the European patent is granted. The Community Patent Convention, which would introduce the first Europe-wide patent, contains provisions for a Community Common Appeal Court. It is easy to imagine the advantages of a unitary patent system: the management of rights would be greatly facilitated and the system would avoid the need for infringement actions to be brought in each Member State.[39] However, there are apparent weaknesses within the Convention. The most important of these is the judicial arrangement set in place. The Convention provides for two procedures that can result in revocation of a Community patent. First, an application for revocation may be filed direct with the EPO. If the grounds for revocation are met the patent is revoked with effect throughout the Community. Second, if, in an action for infringement before a national court, there is a counterclaim for revocation, on any of the grounds laid down in the Convention, and these are met, the national court can revoke the patent. Notwithstanding the possibility of bringing the matter before the Common Appeal Court in such cases, the powers which are conferred by the Convention on a single national court to order the revocation of a Community patent is potentially a source of legal uncertainty.

In the follow-up to the *Green Paper on the Community Patent and the Patent System in Europe*[40] the Commission has defined its choice of measures and worked closely and fruitfully with the EPO.[41] The suggestions outlined in the Green Paper have mostly been adhered to. The European Parliament adopted its opinion on the matter on 19 November 1998. It considers that consistent and effective Community legislation in the field of patents is a vital factor in promoting the competitiveness of

---

37   COM(97) 314 final, 24 Jun 1997.

38   OJ No L 401, 30 Dec 1989, 1. The Agreement is designed to contribute to the achievement of the aims of the single market.

39   Other advantages would include heightening the functioning of the European Internal Market and increasing the profitability of European industry. See Anthony Arnull and Robin Jacob, 'European Patent Litigation: Out of the Impasse?', 2007 vol. 29 *European Intellectual Property Review* 209.

40   Communication from the Commission to the Council, the European Parliament and the Economic and Social Committee, COM/99/0042/final.

41   Ibid. at 7.

enterprises in the EU. For this reason, it concludes, that it is now no longer sufficient to harmonize the concrete provisions of national patent legislation and it is necessary to draw up a Community Regulation which will be valid across the EU.

Meanwhile, work on the modernization of the European patent system moved a step closer when the Administrative Council of the European Patent Organization adopted a Basic Proposal for the Revision of the European Patent Convention on 7 September 2000. The European Patent Organization was given the mandate to hold this conference by the Intergovernmental Conference of its Member States which took place in June 1999. At the second such conference, in October 2000, the same Member States took a number of significant steps towards introducing a common court system for European patent litigation. The underlying aim of the revision conference is to modernize the European system while at the same time retaining fundamental features. In this regard, a number of issues which are of particular relevance to the patent system but which have not yet been debated in full were not on the Conference agenda,[42] including patent protection for biotechnological inventions. It was felt in view of the existing Biotechnology Directive and the fact that the EPC contains identical provisions, there is currently no urgent need for action on this matter. Nor did the Conference feel the need to discuss changes to the EPC arising from the future Community patent.

It can be seen that progress towards the Community Patent was slow. A proposal for a Regulation on the Community Patent was put forward by the Commission.[43] But the complex issues addressed in the Regulation proved difficult to solve quickly. In a somewhat surprising move the Council of Ministers' Competitiveness Committee reached agreement on a 'common political approach' in March 2003 and published a revised text of the proposal for a Council Regulation on the Community Patent.[44] Some of the more important points include:

- The EPO will play a key role in the administration of Community patents and will alone be responsible for examination of applications and the grant of such patents.
- National patent offices will have an important role as set out in the Common approach of 31 May 2001 under which applications for Community patents can be filed with the National Patent Office.
- The accession of the EU to the EPC will enable the EU to be included in the Convention system as a territory for which a unitary patent can be granted.
- Community patent law applicable to the Community patent should not replace the laws of the Member States on patents, nor European patent law as established by the EPC.
- The substantive law applicable to the Community patent, as regards,

---

42 The conference took place in Munich in Nov 2000.
43 COM(2000) 412 final (2000/0177CNS).
44 16 Apr 2003, Interinstitutional File: 2000/0177 (CNS), 8539/03.

patentability, extent of protection, limitation of the effects of the patent, and exhaustion of rights, must follow the same principles as the existing Community legislation with respect to national patents.

- All legal actions relating to certain aspects of the Community patent should come under the jurisdiction of one court whose decisions are enforceable throughout the Community. Actions relating to infringement and validity should be given to the ECJ. jurisdiction shall reside in the first instance in the Community Patent Court and, on appeal, in the Court of First Instance. The ECJ may make a decision in the last resort.

Notwithstanding best efforts for a Council Regulation on the Community Patent, nothing concrete has yet emerged. In February 2007 EU Commissioner Charlie McCreevy said: 'The proposal for an EU-wide patent is stuck in the mud. It is clear to me from discussions with Member States that there is no consensus at present on how to improve the situation.'

Despite setbacks, the Commission continues to recognize the importance of the link between patents and innovation as being an important contributor to competitiveness (outlined in the report *Enhancing the Patent System in Europe*).[45] The Commission accepts the reality that the single market for patents is still incomplete; the creation of a single and affordable[46] Community-wide patent not yet in existence. The fragmented single market for patents has serious consequences for the competitiveness of Europe in relation to the challenges of the US and the emerging economic powers. The Commission believes that in a global competitive economy it is not sustainable for the EU to lose out in an area as crucial for innovation as patent policy. The Commission supports the idea of the Community patent as the approach which will yield most added-value for European industry.

The strategy to promote growth and jobs depends on a close partnership between Member States and the Community. This strategy is outlined in the *Proposal for a Community Lisbon Programme 2008–2010*[47] Some of the 'key' objectives include the Community making a reality of the 'fifth freedom' – the free movement of knowledge; creating a genuine European research area; and improving the framework and conditions for innovation. This will involve the creation of a Community patent and the improvement of the patent litigation

---

45   Communication from the Commission to the European Parliament, the Council, the Economic and Social Committee and the Committee of the Regions, COM (2007) 165 final.

46   In this regard, the London Agreement exists, 17 Oct 2000, and aims to reduce the cost of translations for the European patent. Ten EPC contracting states signed the agreement.

47   Communication from the Commission to the European Parliament, the Council, the Economic and Social Committee and the Committee of the Regions, COM (2007) 804 final.

system, substantially bringing down the cost of patent registration and increasing legal certainty.

In relation to improving the patent litigation system, presently a working document exists and aims at setting out in more technical detail the possible features of a future unified and integrated patent litigation system.[48] The Document discusses areas such as EU patent jurisdiction; main features of the court system; and remedies, procedures and other measures. While there seems to be broad agreement on the overall structure, there are issues that require further discussion. These include, in particular, the composition of panels at first instance, the language of proceedings, the jurisdiction for counterclaims for invalidity, translational arrangements, the modalities for granting the Court of Justice the power to review judgments handed down by the appeal instance, and the funding of the system.[49]

Despite slow progress, developments in the EU are significant and relate to the continued attempt by the EPO to deal with all issues regarding biotechnological inventions. In addition, case law developments are important and continue to evolve in areas such as:

- stem cell cultures
- patents for plant production
- patents for human genes.

Meantime, opinions on patents in the field of biotechnology continue to be divided, with unfettered scientific progress at one end of the spectrum and the basic values accepted by society and the general public, on the other. Given the absence of full and frank discussion in the area, it is unlikely this situation will change in the near future.

Conclusions

The complexity of modern biotechnology gives rise to concerns unprecedented in the modern world and is the reason why morality is now an issue in patent law despite not being particularly so in the past. However, morality as a criterion of patent law was never considered when states adopted the Strasbourg Convention 1963 and the European Patent Convention 1973, suggesting that moral considerations are residual in the patent system. European Patent Office jurisprudence suggests that the policies underlying the creation of Article 53 EPC are unclear and no longer protect adequately biotechnological inventions. Traditional substantive patent law criteria already pose sufficient difficulties for

---

48   Council of the European Union, 27 Feb 2008, 7001/08. The Working Party on Intellectual Property (Patents).

49   Council of the European Union, Progress Report from the Presidency to the Council (Competitiveness), *Enhancing the Patent System in Europe*, 22 May 2008, 9473/08, PI 2.

biotechnological inventions, suggesting that further problems, created in addition by moral constraints on the margins, are undesirable. Because the Biotechnology Directive continues the shortcomings of Article 53 EPC, what the Directive adds in terms of protection is uncertain. The experience of the US offers a comparative perspective and shows, first, that issues raised as moral concerns within patent law are now otherwise addressed, and second, how Europe might proceed, or have proceeded better. However, in view of the difficulties in reaching agreement on the Regulation for the Community Patent, other agreements, outside of the EU legal framework, may be proposed, suggesting the policy of the European Commission is seriously compromised.

Bibliography

Books

Amani, Bita, *State Agency and the Patenting of Life in International Law* (2009) Ashgate Publishing, UK.

Areen, Judith; King, Patricia A; Goldberg, Steven; Gostin, Lawrence, and Capron, Alexander Morgan, *Law, Science and Medicine* 2nd edn (1996) Foundation Press, New York.

Armitage, Edward, and Davies, Ivor, *Patents and Morality in Perspective* (1993) The Intellectual Property Institute, London.

Austin, Martin, *Business Development for the Biotechnology and Pharmaceutical Industry* (2008) Gower Publications, UK.

Bent, Stephen A; Schwaab, Richard L; Conlin, David G, and Jeffery, Donald D, *Intellectual Property Rights in Biotechnology Worldwide* (1987) Stockton Press, New York, and Macmillan Publishers, London.

Beyleveld, Deryck, and Brownsword, Roger, *Mice, Morality and Patents* (1993) The Common Law Institute of Intellectual Property, London.

Chisum, Donald S, *Patents: A Treatise on the Law of Patentability, Validity and Infringement* (1978–) Matthew Bender and Co., New York. Updated periodically, most recently Sept 2000.

Chisum, Donald S; Nard, Craig Allen; Schwartz, Herbert F; Newman, Pauline, and Kiefe, F Scott, *Principles of Patent Law: Cases and Materials* (1998) Foundation Press, New York.

Cornish, W R, *Intellectual Property: Patents, Copyright, Trade Marks and Allied Rights* 4th edn (1999) Sweet & Maxwell, London.

Crespi, R Stephen, *Patents: A Basic Guide to Patenting in Biotechnology* (1988) Cambridge University Press, UK.

Elger, Bernice; Biller-Andorno, Nikola; Mauron, Alexandre, and Capron, Alexander M (eds), *Ethical Issues in Governing Biobanks* (2008) Ashgate Publishing, UK.

Feldman, Robin, *The Internationalisation of Science in Modern Law* (2009) Oxford University Press.

Gibson, Johanna (ed.), *Patenting Lives, Life Patents, Culture and Development* (2008) Ashgate Publishing, UK.

Glasner, Peter, and Rothman, Harry (eds), *Genetic Imaginations: Ethical, Legal and Social Issues in Human Genome Research* (1998) Ashgate Publishing, UK.

Grubb, Philip W, *Patents for Chemicals, Pharmaceuticals and Biotechnology: Fundamentals of Global Law, Practice and Strategy* (1999) Oxford University Press.

Guellec, Dominique, and van Pottelsberghe de la Potterie, Bruno, *The Economics of the European Patent System* (2007) Oxford University Press, UK.

Harris, John, *Wonderwoman and Superman: The Ethics of Human Biotechnology* (1993) Oxford University Press, UK.

Harris, John, *Clones, Genes, and Immortality* (1998) Oxford University Press, UK.

Herring, Jonathan, *Medical Law and Ethics* 2nd edn (2008) Oxford University Press.

Hubbard, Ruth, and Wald, Elijah, *Exploding the Gene Myth* (1997) Beacon Press, Mass.

Jacobs, Alan (ed.), *Patents throughout the World, Part 1 and Part 2*, updated periodically by Elizabeth Hanellin (1996) Clark, Boardman and Callaghan Publishers, New York.

Kitcher, Philip, *The Lives to Come: The Genetic Revolution and Genetic Possibilities* (1996) Penguin, UK.

Lenk, Christian; Hoppe, Nils, and Andorno, Robert (eds), *Ethics and Law of Intellectual Property* (2007) Ashgate Publishing, UK.

Lesser, William (ed.), *Animal Patents: The Legal, Economic and Social Issues* (1989) Macmillian Publishers, London.

Merges, Robert Patrick, *Patent Law and Policy–cases and Materials*, 2nd edn (1997) Michie Law Publishers, Va.

Miller, Arthur R, and Davis, Michael H, *Intellectual Property: Patents, Trademarks and Copyright*, 2nd edn (1990) West Publishing Company, St Paul, Minn.

Mosier, Nathan S, and Ladisch, Michael R, *Modern Biotechnology-Connecting Innovations in Microbiology and Biochemistry to Engineering Fundamentals* (2009) Wiley & Sons.

Oudemans, G, *The Draft European Patent Convention: A Commentary with English and French Texts* (1963) Stevens & Sons, London.

Phillips, Jeremy (ed.), *Patents in Perspective* (1985) ESC Publishing, Oxford.

Reiss, Michael, and Straughan, Roger, *Improving Nature?: The Science and Ethics of Genetic Engineering* (1996) Cambridge University Press, UK.

Rosenberg, Peter D, *Patent Law Fundamentals* vols 1, 2 and 3, 2nd edn (1996) Clark, Boardman and Callaghan Publishers, New York.

Shiva, Vandana, *Biopiracy: The Plunder of Nature and Knowledge (*1998) Green Books Ltd, UK.

Sterckx, Sigrid (ed.), *Biotechnology, Patents and Morality* (1997) Ashgate Publishing, UK.

Stewart Jr., C Neal (ed.), *Plant Biotechnology and Genetics* (2008) Wiley & Sons.

Suzuki, David, and Knudtson, Peter, *Genetics: The Ethics of Engineering Life* (1989) Unwin Hyman Ltd, UK.

Van de Graaf, E S, *Patent Law and Modern Biotechnology* (1997) Gouda Quint, The Netherlands.

Van Empel, M, *The Granting of European Patents* (1974) ESC Publishing, Oxford.

Wheale, Peter; von Schomberg, Rene, and Glasner, Peter (eds), *The Social Management of Genetic Engineering* (1998) Ashgate Publishing, UK.

Essays in Journals and Edited Books

Adams, John N, 'Supplementary Protection Certificates: The "Salt" Problem', 1995 vol. 6 *European Intellectual Property Review* 277–80.

Adcock, Mike, and Llewelyn Margaret, 'TRIPS and the Patentability of Micro-organisms', 2000/2001 vol. 4 no. 3 *Bio-Science Law Review* 91–101.

Antons, Christoph, '*Sui Generis* Protection for Plant Varieties and Traditional Agricultural Knowledge', 2007 vol. 29 *European Intellectual Property Review* 480.

Armitage, Edward, 'EU Industrial Property Policy: Priority for Patents?', 1996 vol. 10 *European Intellectual Property Review* 555–8.

Armitage, Robert A, 'The Emerging US Patent Law for the Protection of Biotechnology Research Results', 1989 vol. 11 *European Intellectual Property Review* 47–57.

Armstrong, George, 'From the Fetishism of Commodities to the Regulated Market: The Rise and Decline of Property', 1987 vol. 82 *Northwestern University Law Review* 79–94.

Arnull, Anthony, and Jacob, Robin, 'European Patent Litigation: Out of the Impasse?' 2007 vol. 29 *European Intellectual Property Review* 209.

Aubrey, J M, 'A Justification of the Patent System', in Jeremy Phillips (ed.), *Patents in Perspective* (1985) 1–9, ESC Publishing, Oxford, UK.

Bagley, Margo A, 'Patent First, Ask Questions Later: Morality and Biotechnology in Patent Law', 2003 45 *Wm. & Mary Law Review* 469.

Bassett, Richard, 'How Will the Directive Affect Patenting Procedures?', *ESC Conference, London, June 1986.*

Bently, Lionel, and Sherman, Brad, 'The Ethics of Patenting: Towards a Transgenic Patent System', 1995 vol. 3 *Medical Law Review* 275–91.

Bently, Lionel, 'Imitations and Immorality: The Onco-Mouse Decision', 1992 vol. 3 *Kings College Law Journal* 145–9.

Berns, A, 1991 vol. 1 *Current Biology*, 28.

Berson, Bennett, 'The Taking of Human Tissue for Research and Commerce: A Comparison of US and French Approaches', 1992 vol. 10 *Wisconsin International Law Journal* 348–70.

Beyleveld, Deryck, 'Why Recital 26 of the E.C. Directive on the Legal Protection of Biotechnological Inventions Should Be Implemented in National Law', 2000 No. 1 *Intellectual Property Quarterly* 1–26.

Beyleveld, Deryck, and Brownsword, Roger, 'Human Dignity, Human Rights and Human Genetics', 1998 vol. 61 *Modern Law Review* 661–80.

Black, Julia, 'Regulation as Facilitation: Negotiating the Genetic Revolution', 1998 vol. 61 *Modern Law Review* 621–60.

Blakeney, Michael, 'The Role of Competition in Bio-technological Patenting and Innovation', 2006/2007 vol. 9 *Bio-Science Law Review*.

Bonadio, Enrico, 'Crop Breeding and Intellectual Property in the Global Village', 2007 vol. 29 *European Intellectual Property Review*.

Bostyn, Sven J R, 'The Patentability of Genetic Information Carriers', 1999 no. 1 *Intellectual Property Quarterly* 1–36.

Braun, Richard, 'Patients' Organisations Support European Union Patenting Rules', 1996 vol. 14 *Trends in Biotechnology* 333–4.

Brom, F W A; Vorstenbosch, J M G, and Schroten, E, 'Public Policy and Transgenic Animals: Case-by-case Assessment as a Moral Learning Process', in P Wheale, R von Schomberg and P Glasner (eds), *The Social Management of Genetic Engineering* (1998) Ashgate Publishing, UK.

Brownsword, R; Cornish, W R, and Llewelyn M (eds), Editors comment: 'Human Genetics and the Law: Regulating a Revolution', 1998 vol. 61 *Modern Law Review* 593–7.

Byrne, Noel J, 'Plants, Animals and Industrial Patents', 1985 vol. 16 *International Review of Industrial Property and Copyright Law* 1–18.

Campbell, Colin, 'A Commission for the 21st Century', 1998 vol. 61 *Modern Law Review* 598–602.

Cohen, Laurence J, 'Litigation in the Biotechnology Industry', *ESC Conference, London, June 1996*.

Correa, Carlos M, 'Some Assumptions on Patent Law and Pharmaceutical R&D', 2001 Quaker United Nations Office, Geneva, Switzerland.

Correa, Carlos M, 'Traditional Knowledge and Intellectual Property', 2001 Quaker United Nations Office, Geneva, Switzerland.

Correa, Carlos M, 'TRIPS Disputes: Implications for the Pharmaceutical Sector', 2001 Quaker United Nations Office, Geneva, Switzerland.

Crespi, R Stephen, 'Prospects for International Cooperation', in William Lesser (ed.), *Animal Patents: The Legal, Economic and Social Issues* (1989) 30–38 Macmillan Publishing, UK.

Crespi, R Stephen, 'Biotechnology, Broad Claims and the EPC', 1995 vol. 6 *European Intellectual Property Review* 267–8.

Crespi, R Stephen, 'Biotechnology Patenting: The Wicked Animal Must Defend Itself', 1995 vol. 9 *European Intellectual Property Review* 431–41.

Crespi, R Stephen, 'Why Does Genetic Material Need Protection and What Are the Benefits and Dangers Allowing Rights over Living Animals?', *ESC Conference, London, June 1996*.

Croyle, Maria A, 'Gene Therapy', in Daan J A Crommelin, Robert D Sindelar, and Bernd Meibohm (eds), *Pharmaceutical Biotechnology: Fundamentals and Applications* 3rd edn (2008) Informa Healthcare, USA Inc.

Czmus, Akim F, 'Biotechnology Protection in Japan, the European Community, and the United States', 1994 vol. 8 *Temple International and Comparative Law Journal* 435–63.

Deech, Ruth, 'Family Law and Genetics', 1998 vol. 61 *Modern Law Review* 697–715.

Drahos, Peter, 'Biotechnology Patents, Markets and Morality', 1999 *European Intellectual Property Review* 441–9.

Ducor, Phillippe, 'Are Patents and Research Compatible?', 1997 vol. 387 (May) *Nature* 13–14.

Dworkin, Gerald, and Kennedy Ian, 'Human Tissue: Rights in the Body and Its Parts', 1993 vol. 1 *Medical Law Review* 291–319.

Ebbink, Richard, 'The Performance of Biotech Patents in the National Courts of Europe', 1995 vol. 75 *Patent World* 25–8.

Ecker, Joseph R, 'Genome Sequencing: Genes Blossom From a Weed', 1998 vol. 391 (Jan) *Nature* 438.

Edgington, Stephen M, 'Plant Patents Double Biotechnology Litigation', 1997 vol. 15 (Mar) *Nature Biotechnology* 216–17.

Editorial, 'Why the European Parliament Rejected the Patenting of Human Genes', 1995 vol. 2 *Gene Therapy* 509–11.

Editorial, 'FDA Needs to Regulate Genetic "Home Brews"', 1996 vol. 14 (Dec) *Nature Biotechnology* 1627.

'The Genome Rush of 1996', 1996 vol. 14 (Oct) *Nature Biotechnology* 1199.

Editorial, 'How Will Biotechnology Fare in Pharma's Big Plan?', 1996 vol. 14 (Nov) *Nature Biotechnology* 1515.

Editorial, 'Biotech's Latest Wrinkle', 1997 vol. 15 (Mar) *Nature Biotechnology* 193.

Editorial, 'Gene Therapy: The US FTC Explains It All to You', 1997 vol. 15 (Feb) *Nature Biotechnology* 109.

Editorial, 'Thinking about Cloning', 1997 vol. 15 (Apr) *Nature Biotechnology* 293.

Editorial, 'Time To Go Public', 14 Nov 1998 vol. 160 *New Scientist* 3.

Editorial, 'Human Genetics and the Law: Regulating a Revolution', 1998 vol. 61 *Modern Law Review* 593–7.

Editorial, 'Vacuum at the Heart of Europe', 1999 vol. 402 (Nov) *Nature* 1.

Farmer, Stacey and Grund, Martin, 'An Overview of the New European Patent Convention (EPC 2000) and Its Potential Impact on European Patent Practice', 2006/2007 vol. 9 no. 2 *Bio-Science Law Review* 53.

Ferry, Georgina, 'The Human Worm', 5 Dec 1998 vol. 160 *New Scientist* 33–5.

Fitt, Robert, and Noddler, Edward, 'Specific, Substantial and Credible? A New Test for Gene Patents', 2008 vol. 9 *Bio-Science Law Review*.

Floyd, Christopher, 'Novelty under the Patents Act 1977: The State of the Art after Merrell Dow', 1996 vol. 9 *European Intellectual Property Review* 480–86.

Ford, Richard, 'The Morality of Biotechnology Patents: Differing Legal Obligations in Europe', 1997 vol. 6 *European Intellectual Property Review* 315–18.

Franzosi, Mario, 'Patentable Inventions: Technical and Social Phases: Industrial Character and Utility', 1997 vol. 5 *European Intellectual Property Review* 251–4.

Friedrich, Glenn A, 'Moving Beyond the Genome Projects', 1997 vol. 14 *Nature Biotechnology* 1234–6.

Garforth, Kathryn, 'Life as Chemistry or Life as Biology? An Ethic of Patents on Genetically Modified Organisms', in Johanna Gibson (ed.), *Patenting Lives: Life Patents, Culture and Development* (2008) Ashgate Publishing, UK.

Gilbert, Penny, and Wilson, Alex, 'Broad Biotech Patents in the United Kingdom after Amgen', 2000/2001 vol. 4 no. 3 *Bio-Science Law Review* 108–12.

Glaser, Vicki, 'Novartis Licenses GENE Therapy to Avoid Monopoly', 1997 vol. 15 (Feb) *Nature Biotechnology* 118–19.

Golub, Edward S, 'Genetically Enhanced Food for Thought', 1997 vol. 15 (Feb) *Nature Biotechnology* 112.

Greengrass, Barry, 'The 1991 Act of the UPOV Convention', 1991 vol. 12 *European Intellectual Property Review* 466–72.

Grund, Martin, 'Are Transgenic Plants and Animals Now Patentable under the European Patent Convention?', 1999/2000 vol. 2 *Bio-Science Law Review* 51–5.

Harms, Derek, 'Drafting Claims around Morality', 1996 vol. 7 *European Intellectual Property Review* 424–5.

Harris, Bob, 'Of Mice and Men', 13 Feb 1999 *New Scientist* 4.

Helmoth, Laura, 'One Mouse's Meat Is Another One's Poison', 1999 vol. 285 (Aug) *Science* 1190–91.

Henderson, Elizabeth, 'TRIPS and the Third World: The Example of Pharmaceutical Patents in India', 1997 vol. 11 *European Intellectual Property Review* 651–63.

Herdegen, Matthias, 'Patenting Human Genes and Other Parts of the Human Body under the EC Biotechnology Directive', 2000/2001 vol. 4 no. 3 *Bio-Science Law Review* 102–7.

Hird, Sean, and Peeters, Michael, 'UK Protection for Recombinant DNA: Exploring the Options', 1991 vol. 9 *European Intellectual Property Review* 334–9.

Hoffmaster, Barry, 'Between the Sacred and the Profane: Bodies, Property, and Patents in the Moore Case', 1992 vol. 7 *Intellectual Property Journal* 115–48.

Hoffmaster, Barry, 'The Ethics of Patenting Higher Life Forms', 1998 vol. 4 *Intellectual Property Journal* 1–24.

Jacob, The Hon. Mr. Justice, 'The Harmonisation of Patent Law', 1997 vol. 2 *Intellectual Property Quarterly* 142–50.

Jones, Nigel, 'The New Draft Biotechnology Directive', 1996 vol. 6 *European Intellectual Property Review* 363–5.

Jones, Nigel, 'Regulation and Patenting: European Harmonisation', 1998/1999 vol. 6 *Bio-Science Law Review* 203–10.

Jones, Nigel, and Britton, Isabel, 'Biotech Patents: The Trend Reversed Again', 1996 vol. 3 *European Intellectual Property Review* 171–3.

Karet, Ian, 'Priority and Sufficiency, Inventions and Obviousness', 1995 vol. 1 *European Intellectual Property Review* 42–6.

Karet, Ian, 'Over-broad Patent Claims: An Inventive Step by the EPO', 1996 vol. 10 *European Intellectual Property Review* 561–3.

Karet, Ian, 'Delivering the Goods?', 1997 vol. 1 *European Intellectual Property Review* 21–31.

Kariyawasam, Kanchana, 'Terminator Technology as a Technological Means of Forcing Intellectual Property Rights in Plant Gernplasm: Its Implications for World Agriculture', 2009 vol. 31 *European Intellectual Property Review*.

Kelly, Helen, 'Biogen's Hepatitis B Patent Held Valid and Infringed', 1994 vol. 2 *European Intellectual Property Review* 75–7.

Kinderlerer, Julian, and Longley, Diane, 'Human Generics the New Panacea?', 1998 vol. 61 *Modern Law Review* 603–20.

Kjeldgaard, Richard H, and Marsh, David R, 'Recent Developments in the Protection of Plant–based Technology in the United States', 1997 vol. 1 *European Intellectual Property Review* 16–20.

Koeing, Robert, 'EU Grabs Food Safety by the Horns', 2000 vol. 287 (Jan) *Science* 403–5.

Kronz, Hermann, 'Patent Protection for Innovations: A Model – Part 1', 1983 vol. 7 *European Intellectual Property Review* 178–83.

Laurie, Graeme, 'Biotechnology and Intellectual Property: A Marriage of Inconvenience?' in Sheila McLean (ed.), *Contemporary Issues in Law, Medicine and Ethics* (1996) 237–67, Ashgate Publishing, UK.

Laurie, Graeme, 'Biotechnology: Facing the Problems of Patent Law', in *Innovation, Incentive and Reward*, Hume Papers on Public Policy vol. 5 (1997) 46–63 Edinburgh University Press.

Laurie, Graeme, '(Intellectual) Property? Let's Think about Staking a Claim to Our Own Genetic Samples', *Intellectual Property and Related Socio-legal Aspects of the Human Genome Project Conference, Edinburgh University, Apr 2001*.

Lawson, Charles, 'Patents and Biological Diversity Convention, Destruction and Decline', 2006 vol. 28 *European Intellectual Property Review*.

Levine, Howard W, 'The Doctrine of Equivalents', 1997 vol. 15 (Apr) *Nature Biotechnology* 383–4.

Llewelyn, Margaret, 'Industrial Applicability/ Utility and Genetic Engineering: Current Practices in Europe and the United States', 1994 vol. 11 *European Intellectual Property Review* 473–80.

Llewelyn, Margaret, 'Article 53 Revisited', 1995 vol. 10 *European Intellectual Property Review* 506–11.

Llewelyn, Margaret, 'The Legal Protection of Biotechnological Inventions: An Alternative Approach', 1997 vol. 3 *European Intellectual Property Review* 115–27.

Llewelyn, Margaret, 'European Plant Variety Protection: A Reactionary Time', 1998/1999 vol. 6 *Bio-Science Law Review* 211–19.

Llewelyn, Margaret, 'The Legal Protection of Biological Material in the New Millennium: The Dawn of a New Era or 21st-century Blues?', 1999/2000 vol. 4 *Bio-Science Law Review* 123–30.

Llewelyn, Margaret, 'The Patentability of Biological Material: Continuing Contradiction and Confusion', 2000 vol. 22 *European Intellectual Property Review* 191–7.

Londa, Bruce S, and Portal, Gerard, 'Strategies for Pharmaceutical Patent Protection in the United States and Europe', 1995 vol. 71 (Apr) *Patent World* 33–6.

Looney, Barbara, 'Should Genes be Patented? The Gene Patenting Controversy: Legal, Ethical and Policy Foundations of an International Agreement', 1994 vol. 26 *Law and Policy in International Business* 231–72.

MacDonald, Stuart, 'Exploring the Hidden Costs of Patents', 2001 Quaker United Nations Office, Geneva, Switzerland.

MacKenzie, Debora, 'Gut Reaction: Could a Mechanical Gourmet Help Us Digest a GM Future?', 30 Jan 1999 *New Scientist* 4.

MacQuitty, Jonathan J, 'The Real Implications of Dolly', 1997 vol. 15 (Apr) *Nature Biotechnology* 294.

Malinowski, Michael J, 'Globalisation of Biotechnology and the Public Health Challenges Accompanying It', 1996 vol. 60 *Albany Law Review* 119–69.

Markham, A F, 'Introduction to Law and Genetics', 22 Jan 1999 *Conference on the Law and Genetics*, Signet Library, Edinburgh.

Marshall, James M, 'Biogen v Medeva: Lack of Support a Ground for Revocation after All', 1997 (Mar) *Patent World* 34–37.

Marshall, James M, 'Biotechnology Patents: A Further Twist', 1998 (Feb) *Patent World* 25–8.

McInerney, Anthony, 'Biotechnology: Biogen v Medeva in the House of Lords', 1998 issue 1 *European Intellectual Property Review* 14–21.

McLean, Sheila A M, 'Interventions in the Human Genome', 1998 vol. 61 *Modern Law Review* 681–96.

Meibom, Wolfgang von, and Pitz, Johann, 'Broad Biotech Claims–Part 1', 1996 vol. 84 (Aug) *Patent World* 29–32.

Meibom, Wolfgang von, and Pitz, Johann, 'Broad Biotech Claims–Part 2', 1996 (Sept) *Patent World* 30–34.

Meredith, Rick, 'Winning the Race to Invent', 1997 vol. 15 (Mar) *Nature Biotechnology* 283–4.

Merges, Robert Patrick, 'Intellectual Property in Higher Life Forms', 1988 vol. 47 *Maryland Law Review* 1051.

Mitchell, Alison, 'Nuclear Transplantation: The Science of the Lambs', 1 Jan 1998 vol. 391 *Nature* 21.

Morze, Herwig von, and Hanna, Peter, 'Critical and Practical Observations Regarding Pharmaceutical Patent Term Restoration in the European

Communities (Part 1)', 1995 vol. 77 *Journal of Patent and Trademark Office* 479–500.

Mosier, Nathan S, and Ladisch, Michael R, 'Genomes and Genomics', in *Modern Biotechnology: Connecting Innovations in Microbiology and Biochemistry to Engineering Fundamentals* (2009) Wiley & Sons.

Moufang, Rainer, 'Patentability of Genetic Inventions in Animals', 1989 vol. 20 *International Review of Industrial Property and Copyright Law* 823–46.

Moufang, Rainer, 'Methods of Medical Treatment under Patent Law', 1993 vol. 24 *International Review of Industrial Property and Copyright Law* 18–49.

Mueller, Janice M, 'The Evolving Application of the Written Description Requirement to Biotechnological Inventions' 1998 vol. 12 no. 2 *Berkeley Technology Law Journal* 615–52.

Naeem, Shadid, and Li, Shibin, 'Biodiversity Enhances Ecosystems Reliability', 4 Dec 1997 vol. 390 *Nature* 507–509.

Nott, Robin, 'Plants and Animals: Why They Should Be Protected by Patents and Variety Rights', 1993 (July/Aug) *Patent World*.

Nott, Robin, 'The Biotechnology Directive: Does Europe Need a New Draft?', 1995 vol. 12 *European Intellectual Property Review* 563–7.

Nott, Robin, 'The Novartis Case in the EPO', 1999 *European Intellectual Property Review* 33–6.

O'Donnell, Mark, 'Protecting Biotechnology Inventions in Australia', 1998/1999 vol. 6 *Bio-Science Law Review* 220–27.

O'Neill, Onora, 'Insurance and Genetics: The Current State of Play', 1998 vol. 61 *Modern Law Review* 716–23.

Onwuekwe, Chika B., 'Plant Genetic Resources and the Associated Traditional Knowledge: Does the Distinction between Higher and Lower Life Forms Matter?', in Johanna Gibson (ed.), *Patenting Lives: Life Patents, Culture and Development* (2008) Ashgate Publishing, UK.

Oyewunmi, Adojoke, 'The Right to Development, African Countries and the Patenting of Living Organisms: A Human Rights Dilemma', in Johanna Gibson (ed.), *Patenting Lives: Life Patents, Culture and Development* (2008) Ashgate Publishing UK.

Parker, John, 'Reform of the Pharmaceutical Patent Term', 1994 Dec – 1995 Jan vol. 68 *Patent World* 27–32.

Paver, Michelle, 'All Animals Are Patentable, but Some Are More Patentable Than Others', 1992 vol. 9 *Patent World* 9–15.

Peace, Nicholas, and Christie, Andrew, 'Intellectual Property Protection for the Products of Animal Breeding', 1996 vol. 4 *European Intellectual Property Review* 213–33.

Pendleton, Michael, 'Intellectual Property, Information-based Society and a New International Economic Order: The Policy Options?', 1985 vol. 2 *European Intellectual Property Review* 31–4.

Pepa, Stevan M, 'International Trade and Emerging Genetic Revolutionary Regimes', 1998 vol. 29 *Law and Policy in International Business* 415–50.

Phillips, Jeremy, 'The English Patent as a Reward for Invention: The Importation of an Idea', 1983 vol. 2 *European Intellectual Property Review* 41–4.

Pottage, Alain, 'The Inscription of Life in Law: Genes, Patents and Bio-politics', 1998 vol. 61 *Modern Law Review* 740–65.

Purvis, Iain, 'Patents and Genetic Engineering: Does a New Problem Need a New Solution?', 1987 vol. 12 *European Intellectual Property Review* 237–348.

Rai, Arti K, 'Intellectual Property Rights in Biotechnology: Addressing New Technology', 1999 vol. 34 no. 3 *Wake Forest Law Review* 827–47.

Rangnekar, Dwijen, 'Is More Less? An Evolutionary Economics Critique of the Economics of Plant Breeds' Rights', in Johanna Gibson (ed.), *Patenting Lives: Life Patents, Culture and Development* (2008) Asghgate Publishing, UK.

Reid, Brian C, 'Biogen in the EPO: The Advantage of Scientific Understanding', 1995 vol. 2 *European Intellectual Property Review* 98–100.

Reinbothe, Jorg, and Howard, Anthony, 'The State of Play in the Negotiations on TRIPS (GATT/Uruguay Round)', 1991 vol. 5 *European Intellectual Property Review* 157–64.

Roberts, Carol, 'The Prospects of Success of the National Institute of Health's Human Genome Application', 1994 vol. 1 *European Intellectual Property Review* 30–36.

Roberts, Tim, 'Broad Claims for Biotechnological Inventions', 1994 vol. 9 *European Intellectual Property Review* 371–3.

Roberts, Tim, 'Patenting Plants around the World', 1996 vol. 10 *European Intellectual Property Review* 531–6.

Schatz, Ulrich, 'Patents and Morality', in Sigrid Sterckx (ed.), *Biotechnology, Patents and Morality* (1997) 159–70, Ashgate Publishing, UK.

Scott, Andrew, 'The Dutch Challenge to the Bio-patenting Directive', 1999 *European Intellectual Property Review* 212–15.

Scott-Ram, Nick, 'Does Industry Need the Directive?', *ESC Conference, London, June 1996.*

Sherman, Brad, 'Patent Law in a Time of Change: Non-obviousness and Biotechnology', 1990 vol. 10 *Oxford Journal of Legal Studies* 278–87.

Sherman, Brad, 'Patent Claim Interpretatiopn: The Impact of the Protocol on Interpretation', 1991 vol. 54 *Modern Law Review* 499–510.

Smith, Andrew R, 'Monsters at the Patent Office: The Inconsistent Conclusions of Moral Utility and the Controversy of Human Cloning', 2003 vol. 53 *De Paul Law Review* 159.

Spranger, Tade Matthias, 'Ethical Aspects of Patenting Human Genotypes according to EC Biotechnology Directive', 2000 vol. 31 *International Review of Industrial Property and Copyright Law* 373–80.

Sterckx, Sigrid, 'European Patent Law and Biotechnological Inventions', in Sigrid Sterckx (ed.), *Biotechnology, Patents and Morality* (1997) 1–55, Ashgate Publishing, UK.

Straus, Joseph, 'Patent Protection for New Varieties of Plants Produced by Genetic Engineering: Should "Double Protection" Be Prohibited?',1984 vol. 15 *International Review of Industrial Property and Copyright Law* 426–42.

Straus, Joseph, 'Patent Protection for Biotechnological Inventions', 1985 vol. 16 *International Review of Industrial Property and Copyright Law* 445–8.

Straus, Joseph, 'AIPPI and the Protection of Inventions in Plants: Past Developments, Future Perspectives', 1989 vol. 20 *International Review of Industrial Property and Copyright Law* 600–621.

Tansey, Geoff, 'Trade, Intellectual Property, Food and Biodiversity: Key Issues and Options for the 1999 Review of Article 27.3(b) of the TRIPS Agreement', 1999 Quaker United Nations Office, Geneva, Switzerland.

Thomas, Sandra, 'Intellectual Property Rights in Biotechnology' 1996 vol. 6 *Therapeutics Patents* 845–54.

Thomas, Sandra M; Kimura, K, and Burke, J F, 'Patenting of Recombinant Proteins: An Analysis of Tissue Plasminogen Activator in Europe, the United States, and Japan', 1995 vol. 24 *Research Policy* 645–63.

Thomson, Judith, 'The Grey Penumbra of Interpretation Surrounding the Obviousness Test for Biotech Patents', 1996 vol. 2 *European Intellectual Property Review* 90–96.

Thurston, Julian, 'Self-replication Biological Material and the Law of Property', 1994 (Apr), *McKenna & Co*, London.

Thurston, Julian, 'How to Effectively License Genetic Material', *ESC Conference, London, June 1996.*

Uhlen, Mathias, 'Whose Genome Is It Anyway', 1995 vol. 13 *Trends in Biotechnology* 160–62.

Vandergheynst, Dominique, 'The Revised Draft on the Legal Protection of Biotechnological Inventions', 1996 Jun, *ESC Conference*, London.

Van Overwalle, Geertrui, 'Protecting Innovations in Plant Biotechnology: Patents or Plant Breeders' Rights?', 24–25 Sept 1992 *Sixth Forum for Applied Biotechnology* (also in special issue (Mar 1993) of *Mededelingen van de Faculteit Landbouwwetenschappen*, University of Gent).

Van Overwalle, Geertrui, 'The Legal Protection of Biotechnological Inventions in Europe', 29 Feb 1996 *Generale Bank Chair* 1–66 (also in *Intellectual Property Rights and Strategic Alliances in the European Union,* Leuven Law Series (1997) Leuven University Press).

Van Overwalle, Geertrui, 'Biotechnology Patents in Europe: From Law to Ethics', in Sigrid Sterckx (ed.), *Biotechnology, Patents and Morality* (1997) 138–9, Ashgate Publishing, UK.

Van Overwalle, Geertrui, 'Patent Protection for Plants: A Comparison of American and European Approaches', 1999 vol. 39 no. 2 *The Journal of Law and Technology* 143–94.

Ventose, Eddy D, 'Farming Out an Exemption for Animals to the Method of Medical Treatment Exclusion under the European Patent Convention, 2008 vol. 30 *European Intellectual Property Review.*

Vogel, Gretchen, 'Company Gets Rights to Cloned Human Embryos', 2000 vol. 287 (Jan) *Science* 559.

Weiss, Rick, 'Patent Sought on Making of Part-human Creatures', *Washington Post*, 2 April 1998, A 12.

Wells, Celia, '"I Blame the Parents": Fitting New Genes in Old Criminal Laws', 1998 vol. 61 *Modern Law Review* 724–39

Wells, Angus J, 'Patenting New Life Forms: An Ecological Perspective', 1994 vol. 3 *European Intellectual Property Review* 111–18.

Whaite, Robin, and Jones, Nigel, 'Pharmaceutical Patent Term Restoration: The European Commission's Proposed Regulation' 1992 (Sept) *European Intellectual Property Review* 324–6.

White, Alan W, 'Patentability of Medical Treatment: Wellcome Foundation's (Hitching's) Application', 1980 (Nov) *European Intellectual Property Review* 364–9.

White, Alan W, 'Patenting the Second Medical Indication', 1985 vol. 3 *European Intellectual Property Review* 62–9.

White, Alan W, and Brown John D, 'EPC Appeal Procedures', 1996 vol. 7 *European Intellectual Property Review* 419–23.

Woessner, Warren D, 'Patenting Transgenic Animals: From the Harvard Mouse to "Hello Dolly"', 1999 vol. 1 *Bio-Science Law Review* 3–7.

Wright, Jacqueline D, 'Implications of Recent Patent Law Changes on Biotechnology Research and the Biotechnology Industry', 1997 vol. 1 *Virginia Journal of Law and Technology* 1–13.

Reports and Other Official Publications

Advisory Commission on Patent Law Reform (Aug 1992) 1–27. Report to the Secretary of Commerce the Honourable Barbara Hackman Franklin, USA.

Amended Proposal for a Council Directive on the Legal Protection of Biotechnological Inventions, OJ (EC), 93/C 44/03, 16 Feb 1993.

Annas, J G, *Of Monkeys, Man and Oysters* (1987), Hastings Center Report No. 4, 20.

Banks, M A L, *The British Patent System, Report of the Committee to Examine the Patent System and Patent Law* (1970: Cmnd 4407). HMSO.

Byrne, Noel J, *The Scope of Intellectual Property Protection for Plants and Other Life Forms* (1989) Intellectual Property Publishing Ltd, UK.

Commission Proposal for Directive of 21 Oct 1988 [COM (88) 0496 final].

Commission of the European Communities, *Communication from the Commission to the European Parliament, the Council, the European Economic and Social Committee and the Committee of the Regions*, 11 Dec 2007, COM (2007) 804 final.

Commission of the European Communities, *Communication from the Commission to the European Parliament and the Council: Enhancing the Patent System in Europe*, 4 Apr 2007, 8302/07.

Council of the European Union, *EU Patent Jurisdiction: Main Features of the Court System; Remedies, Procedures and Other Measures*, 27 Feb 2008, 7001/08.

Council of the European Union, *Enhancing the Patent System in Europe*, 22 May 2008, 9473/08.

Curry, Judith R, *The Patentability of Genetically Engineered Plants and Animals in the US and Europe* (1987) Intellectual Property Publishing Ltd, UK.

Department of the Environment and Local Government of Ireland, *Genetically Modified Organisms and the Environment*, consultation paper, Autumn 1998.

Economic and Social Research Council, *Intellectual Property Research* (1994).

European Commission, Communication from the Commission to the European Parliament, the Council, the Economic and Social Committee.

European Commission, Explanatory Memorandum to the Directive, *Proposal for a European Parliament and Council Directive on the Legal Protection of Biotechnological Inventions*, COM (95) 661.

European Commission, *Green Paper on the Community Patent and the Patent System in Europe*, COM (97) 314 final.

European Commission, Communication to the Council, the European Parliament and the Economic and Social Committee, *Promoting Innovation through Patents*, COM /99/0042 final.

European Commission, 'Promoting the Competitive Environment for the Industrial Activities Based on Biotechnology within the Community', in *The European Group on Ethics in Science and New Technologies* (Dec 1999) (SEC (91) 629 final, 19 Apr 1991).

European Commission, Proposal for Directive, *Proposal for a European Parliament and Council Directive on the Legal Protection of Biotechnological Inventions*, OJ C 296 of 8 Oct 1996 and OJ C 311 of 11 Oct 1997.

European Commission, *Research and Technology: The Fourth Framework Programme (1994–1998)*, Office for Official Publications of the European Communities, Luxembourg, 1995.

European Commission, Communication on Biotechnology, *White Paper on Growth, Competitiveness and Employment* (see 'Preparing the Next Stage', COM (94) 219, 1 June 1994).

European Commission Secretariat-General Directorate C, *The European Group on Ethics in Science and New Technologies* (Dec 1999).

European Parliament, *Report on the Future of the Biotechnology Industry*, 28 Feb 2001, A5–0080/2001.

European Parliament Resolutions, B3–0199, 0220 and 024/93 adopted on 11 Feb 1993, OJ EC 1993, C 72/127.

European Parliament and Council of the European Union, case C–377/98, [1998] OJ C 378/13.

European Patent Office, Comments on the First Preliminary Draft Convention relating to a European Patent Law, 14 Mar 1961, Document IV/2071/61–E.

European Patent Office, Minutes of the Proceedings of the First Meeting of the Patents Working Party, 17–21 Apr 1961, Document IV/2767/61–E.

European Patent Office, Working Party Document IV/2767/61 section 5.

European Union and Council, Amended Proposal for a Directive of the European Parliament and of the Council on the Legal Protection of Biotechnological Inventions, File no. 95/0350 (COD), 19 Nov 1997.

Guidelines for Examination of the European Patent Office (1977) and (1999).

Guidelines for Utility Examination of the USPTO, *Federal Register* vol. 66 no. 4 Jan 2001.

House of Commons Select Committee on Science and Technology, third report, *Human Genetics: The Science and Its Consequences*, HC 1994–95, HC Paper 41.

House of Lords Select Committee on the European Communities, HL Paper 28, 1 Mar 1994, HMSO.

House of Representatives Report 1129, 71st Congress, 2d session, 1930.

House of Representatives Report 1445, 83d Congress, 2d session, 1954.

Human Genetics Commission, UK, 'Whose Hands on Your Genes?', 2000 Lon 15502, Maidenhead SL6 2bz, UK (www.hgc.gov.uk).

Intellectual Property Forum, 'The Way Forward: The Intellectual Property Forum', Mr Justice Jacobs, 26 Apr 1996, London.

Intergovernmental Conference for the Setting Up of a European System for the Grant of Patents, 1970 vol. 1 *International Review of Industrial Property and Copyright Law* 26–31.

International Union for the Protection of New Varieties of Plants, *General Information*, World Intellectual Property Organization publication no. 400(E), 1995.

International Union for the Protection of New Varieties of Plants, *What It Is, and What It Does*, UPOV publication no. 437(E), September 1996.

London Agreement, 17 Oct 2000, published [2001] OJEPO 549.

Nuffield Council on Bioethics 'Ethical Principles: Respect for Human Lives and the Human Body', in *Human Tissue: Ethical and Legal Issues* (Apr 1995).

Office of Technology Assessment Special Report No. 5, *New Developments in Biotechnology*, US Congress (April 1989) Washington DC.

Official Journal of Plant Varieties, 1 July 1997, Dept. of Agriculture, Ireland.

Opinion of the Economic and Social Committee on the 'Proposal for a European Parliament and Council Directive on the Legal Protection of Biotechnological Inventions' OJ (EC), C 295/11, 7 Oct 1996.

Patent Office's Manual of Patent Practice, 1991.

Senate Report no. 315, 71st Congress, 2d session, 1930.

Senate Report no. 1937, 83d Congress 2d session, 1954.

Social and Economic Committee Opinion on the Proposal for Directive, Proposal for a European Parliament and Council Directive on the Legal Protection of Biotechnological Inventions, OJ C 295 of 7 Oct 1996.

Tobin, Brendan, *The Search for Equity in the International Co-operative Biodiversity Group* (ICBG) *Project in Peru*, 30 Sept 1997, Colombia Press.

World Intellectual Property Organization, *Monthly Review*, 1998 Jan no. 1.

Index